Practical Cost Planning

A guide for surveyors and architects

Duncan P. Cartlidge, AIQS, ACIArb
Senior Lecturer in Quantity Surveying at Thames Polytechnic

and Ian N. Mehrtens, BSc., ARICS
Senior Lecturer in Quantity Surveying at Thames Polytechnic

Hutchinson
London Melbourne Sydney Auckland Johannesburg

Hutchinson & Co. (Publishers) Ltd

An imprint of the Hutchinson Publishing Group

17–21 Conway Street, London W1P 5HL

Hutchinson Group (Australia) Pty Ltd
30–32 Cremorne Street, Richmond South, Victoria 3121
PO Box 151, Broadway, New South Wales 2007

Hutchinson Group (NZ) Ltd
32–34 View Road, PO Box 40–086, Glenfield, Auckland 10

Hutchinson Group (SA) (Pty) Ltd
PO Box 337, Bergvlei 2012, South Africa

First published 1982

Set in Times by Bookens, Saffron Walden, Essex

Printed in Great Britain by The Anchor Press Ltd
and bound by Wm Brendon & Son Ltd
both of Tiptree, Essex

British Library Cataloguing in Publication Data

Cartlidge, Duncan P.
  Practical cost planning: a guide
  for surveyors and architects
  1. Buildings – Estimates
  I. Title II. Mehrtens, Ian N.
  692'.5   TH435

ISBN 0 09 146841 8

Contents

Introduction

As teachers of students on degree courses in quantity surveying in the field of cost planning, and as former and current practising surveyors, we have become increasingly aware of the lack of practical-based published literature in this field.

Students often complain that although there is sufficient relevant information in a particular book, the presentation of that information makes it difficult for them to follow the examples which incorporate all the necessary adjustments that invariably have to be made from one step to another, when those examples have differing project details.

It is because of this that we have undertaken to revise the two books previously written by Duncan Cartlidge for Hutchinson, namely *Cost Planning and Building Economics* and *Construction Design Economics*, and to base all examples on one project, hence giving the student a practical insight into what he may be required to do in a live situation.

We trust, therefore, that this book will be of great assistance to all students of cost planning, be they academic or practitioners studying for professional examinations.

Duncan Cartlidge and Ian Mehrtens
1982

1 The need for cost planning and cost control

If it is suggested to a layman that the cost planning and cost control of buildings is a new technique, he will probably reply, in disbelief, that people have been using costing methods for years. Costing is generally accepted to mean applying prices to a schedule of items of labour and materials in order to obtain an approximation of the cost of a construction project. Unfortunately the reputation of the accuracy of the traditional methods is not at all good, a reputation that has spread far beyond the people directly concerned with producing these sometimes aptly-named 'guesstimates'.

During the last few years, however, the traditional estimating methods have been gradually superseded by cost planning techniques.

The controlling of the cost has to be complete and all-embracing if it is to have any contribution in stimulating confidence in the figures that are produced by construction cost consultants.

The effectiveness of cost planning at the pre-tender stages is seriously curtailed if cost control is not exercised for the duration of the contract. The objective of cost planning and cost control is not only to determine the probable cost of a building but to control the design development throughout the project, and to provide the client with value for money.

Cost has a continuous influence on a building project; it will not only dictate the nature of the most important characteristic of the embryo design, the plan shape, storey height, type of finishings and so on, but throughout the design development cost will be exerting its influence.

Even after the building is completed, the effect of cost will be inescapable. Decisions taken during the design sequence will have determined the types of materials that have been incorporated into the structure, those destined for obsolescence within five years, or those with an estimated life at least as long as the usefulness of the building as a whole.

Uneconomical cost distribution at the outset can result in such permanent waves of unsettlement in the design team, resulting in project redesign, costly delays and the like, that time must be spent in the early stages of the design sequence for cost planning to be carried out, and then time should be allocated in order that careful cost control can be operated during site works. The design team must be in a position to monitor accurately the cost of a project from the preliminary stages, to the signing of the final account.

Cost planning, to be at all useful, must perform certain functions. First and perhaps most important, it must provide, early on, usually at the feasibility stage, a reliable preliminary estimate. Without this, then the time spent on producing any preliminary report becomes an academic exercise of little practical use to the person who pays for it all, that is to say the client.

The man who says that the first estimate of cost that is presented to the client cannot hope to be realistic, is undermining one of the chief objectives of cost planning, to produce, before all the detail drawings are prepared in the detail design stage, a preliminary estimate that is reliable. Indeed, it is often considered that the first figure the client hears is the one that he remembers, hence the first estimate of cost must be realistic. Only this will stimulate confidence in the building economist's advice. If it is not possible to provide this service then cost planning can offer no advantages over the wildly inaccurate traditional costing methods, which are so widely scorned.

You might ask yourself, why should the quantity surveyor be the person to carry out the cost control of a project. The simple answer is that it is the quantity surveyor who has the historical background in costing. As far back as the eighteenth century, 'measurers' (as they were called) were employed to measure, value and settle the cost of work after it was both designed and executed. During the nineteenth century, as the main contractor system developed along with price competition, so the measurers saw they had a skill to assist the contractor in measuring and pricing their work. Thus the system developed as far as today's system where an independent surveyor measures the works for the contractor, for competitive pricing by the contractor.

The quantity surveyor's history then is steeped in an air of 'cost'. As resources have become more scarce, modern buildings more complex, and the cost of land and materials more expensive, clients' requirements have been such that they have needed an accurate forecast of the total cost. Who better, then, with his professional history, to cope with the task of cost forecasting and cost control than the quantity surveyor.

Indeed, it has been suggested in recent reports from the Royal Institution of Chartered Surveyors that the next major step for quantity surveyors is into the field of project management. With the emphasis in the industry concentrated on cost and cost control, and with the quantity surveyor's history of cost and inbred management capabilities, then it is a natural progression to project management. Already, one major contract in the British Isles has had as its project manager a leading quantity surveyor Frank E. Graves (a past president of the Royal Institution of Chartered Surveyors). Is this the future role for the quantity surveyor?

Over the years, much research and development by quantity surveyors has been concentrated on this field of cost planning to meet the needs of the industry such that it is now possible to take account not only of the actual building cost but also to consider costs affecting the whole life of buildings in order to have a knowledge of the total cost of building.

The system has developed over the years into a well defined process of cost planning and cost control. As shown in Table 1, cost planning takes place from the *inception stage* of the project to the *scheme design stage*, with

Design sequence	Process	Duty	Method used
1 Inception	Cost planning	Cost range	Interpolation
2 Feasibility	Cost planning	Feasibility study	Interpolation
3 Outline proposals	Cost planning	Confirm cost limit	Single price estimating
4 Scheme design	Cost planning	Cost plan	Single price rate estimating
5 Detail design	Cost control	Cost checking	Approximate quantities
6 Production information	Cost control	Specification	–
7 Bill of quantities	Cost control	BQ	–
8 Tender action	Cost control	Cost analysis	Elemental breakdown of tender

Table 1

cost control from the *detail design stage* to settlement of the final account. The quantity surveyor has duties to perform during this system, but it should be remembered that the system must remain flexible in order for it to work.

Inception

The first stage in the design of any building contract. The client approaches the architect of his choice and is able to provide either one or both of the following pieces of information:

1 User requirements, for example the floor area needed
2 The target cost, for example the total amount of capital available for the project

In the first instance the client would wish to know the probable cost of fulfilling his requirements, before continuing. In the second case the client would wish to know what it would be possible to provide for his capital outlay.

At this stage it is dangerous for the quantity surveyor to give any definite cost estimations. He should simply quote previous known cost analyses, which have been updated, to give the client an idea of the range of costs he is likely to incur for the user requirement stated.

Feasibility

At this stage in the design sequence the architect will have established the nature of the professional assistance he will require for the project. The consultants then appointed produce for the client a *feasibility report*, which is a joint effort of all the professions involved. At this stage the process of designing a building that will exactly conform to the client's wishes becomes a team effort. The design team is formed; the nature of the team will of course depend upon the contract, but a typical line-up could be: architect, quantity surveyor, structural engineer and specialist subcontractors' representatives. Perhaps one of the most important features of designing by team is good communications between members. Anything else can only result in misunderstandings and needless, costly delays.

The feasibility stage is the first time in the design sequence that the client, aided by the design team, will take a major decision, namely whether or not to continue with the project. This will be possible by closely examining the 'pros' and 'cons' of the feasibility report. A report should contain some of the following items:

1   The location of the proposed site
2   The target cost
3. A pictorial image either from photographs of previous projects, or architect's sketches, of the completed design
4   A list of the total floor areas
5   The type of accommodation to be provided
6   The nature of the ground, together with the maximum safe bearing capacities
7   Whether planning permission is available
8   In the case of a factory, the nearest source of skilled and unskilled labour
9   The adequacy of public transport and social facilities
10 The nearest available sources of electricity, gas, water and so on
11 The nearest public sewer

Should it be that the client has not himself imposed a cost target (this becoming more popular in light of the recent economic climate), the quantity surveyor must ascertain a limit. This he can do in a number of ways which will be explored in detail in Chapter 5.

Assuming that the decision is taken to continue, the first major step is now complete, the importance of which cannot be over-emphasized. Now that the cost is known, there must be a method of closely monitoring it throughout the design development and the site works, to ensure that the final account figure does not exceed the target cost. This is achieved by relating the estimate to the actual building; the technique employed is called *elemental cost planning*.

When a tender is being prepared, the cost of a project is calculated by adding together the many hundreds, perhaps thousands, of priced items contained in a bill of quantities. It is possible to itemize accurately all the operations necessary at this stage because of the detailed nature of the information available. However, during the scheme design stage, when the cost plan is prepared, and information is limited, the project is divided into approximately thirty elements. An element is that part of a building described according to the function it performs, irrespective of its actual construction – for example, roof, external walls and staircases.

There are between thirty and thirty-six such elements and sub-elements used in cost planning. It is a sad reflection on the present building economists that to date they find it impossible to agree on a list of elements, although an attempt has now been made to achieve a standard form of cost analysis: see Appendix B. The most commonly used lists are those published by:

1   The Building Cost Information Service (BCIS)
2   The *Architects' Journal*

The service provided by the Royal Institution of Chartered Surveyors Building Cost Information Service is available to members of the RICS and the Institute of Quantity Surveyors. Subscribers are issued at regular intervals with cost analyses and other cost information, in return for which they are expected to contribute cost information to the organization for the benefit of other members.

A *cost analysis* is a record of how the cost has been distributed over the elements of a building; it includes a brief description of the overall nature of the project and specification notes on the general level of finishings, and so on. Therefore, basically, the cost plan is prepared by finding a cost analysis of a similar project to the one under consideration, and by studying how the cost was allocated in the old project, to prepare an estimate for the new scheme.

The cost information cannot be used directly from the cost analysis, and some degree of modification will have to be made. These modifications are usually for differences in:

1   Price level
2   Quantity
3   Finishes and standards

Exactly how this is achieved will be fully described later in this book.

Outline proposals and scheme design

The lump sum estimate must now be split down into the amount of money it is intended to spend on each element (see Figure 1).

This proportioning out is carried out during the *outline proposals* and the *scheme design stages.*

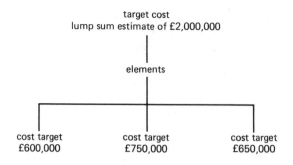

Figure 1

As the project design develops and the architect crystallizes his ideas, the information available to the quantity surveyor is constantly increasing and perhaps changing. At the end of the scheme design stage the architect should have fully developed the client's proposals and after the detail design stage the general layout should not be radically altered. At the outline proposals stage, the information available should have increased to include such items as the outline drawings, and so on.

In theory, armed with this information, the quantity surveyor should be able to prepare an outline cost plan. However, to many design teams this stage is rather an inactive period. The feasibility report is certainly now available, but the outline drawings are very seldom, if ever, prepared. Indeed, for what use they are there is really no reason why they should be.

The constituent parts of a cost analysis are:

1   A summary of project, contract and design/shape information
2   A summary of element costs
3   An amplified analysis

A *brief cost summary*, which is a statement of how the target cost has been distributed over major groups of elements, such as:

1   Substructure
2   Superstructure
3   Internal finishes
4   Furniture and fittings
5   Services
6   External works

is now incorporated within the summary of element costs. It is sometimes found useful, before proportioning the cost over individual elements, to make at the outline proposals stage a broad but accurate cost allocation over these groups of elements.

At the scheme design stage, the architect will produce sketch plans and outline specification notes; from these elements the cost plan is prepared. Also at this stage the architect will have to examine the various methods of meeting the client's requirements, for example how to provide a large uninterrupted floor area for a factory production line, and certainly at some time during this stage the building economist's advice will be sought on the cost implications of various alternative forms of construction, finishings, and so on.

The object of the processes shown in Table 1 is that at any time during the contract the quantity surveyor is in a position accurately to report to the architect:

1 The cost of the building to the client, when completed
2 Whether the general standards of quality being offered are too high or too low for the cost targets

Not only does the proportioning of cost among the elements show how the cost is distributed and whether any element has been overloaded with capital, but it allows the quantity surveyor to alter the balance of cost within the overall target cost.

Consider this situation, remembering all the time that the target cost must not be altered.

Figure 2

For the sake of simplification it is assumed that the target cost has been proportioned among three elements: frame, roof and services.

After the amount set against the frame, element number one, has been carefully checked with the detail design, it becomes obvious that £150,000 is an unrealistic target and that at least an additional £15,000 will be required for this element. To prevent the target cost now rising to £300,000 + £15,000 = £315,000, compensating reductions must be made to the other elements. In the hypothetical situation described it is decided to reduce the cost target of services, element number three. After studying the specification it is decided to substitute a cheaper form of installation. The estimated saving adopting this solution will be £20,000. This sum can now be added to the cost target of element number one. Therefore the revised cost targets read:

Element number 1	frame	£170,000
Element number 2	roof	£ 50,000
Element number 3	services	£ 80,000
	Target cost	£300,000

In performing this 'lending and borrowing' exercise, the quantity surveyor must always remember that he runs the risk of tipping the balance of the design. Care must be taken not to overload one element at the expense of another.

Thus it is argued by some quantity surveyors that the client would much prefer to pay the extra £15,000 than to have the project's overall standards reduced, or the balanced design abolished. The problem then arises, however, of where the line is drawn between acceptable extras of this nature and unacceptable extras, and of the usefulness of the preliminary estimate. Does it merely become an outdated record? If performed correctly there is no reason why the substitution of a less expensive alternative should necessarily result in a lowering of standards, or an abolition of the balanced design.

Detail design

Until this point the process has been one of planning (see Table 1), but from now on, the target cost having been determined, the process becomes one of controlling the cost in order that the tender figure does not exceed the preliminary estimate.

During the detail design stage the architect prepares, for the first time, the detail drawings. As soon as these drawings are completed they are dispatched to the quantity surveyor, who, by using the approximate quantities, cost checks the architect's detail design against the cost target set by it at the preparation of the cost plan. Each element should be cost checked as soon as possible after being detailed, to allow any necessary redesign or reallocation of cost to be carried out. After all the elements have been satisfactorily checked, the architect's drawings are passed to draughtsmen, who produce drawings of the quality that is necessary for the production of a bill of quantities.

2 Factors affecting economic design

This chapter is dedicated to the architect, who, unknowingly, created a lasting monument to uneconomical design shape. The building which is close to Clapham Junction, one can only assume, was originally designed to accommodate the needs of a single client. The solution that the architect produced from the client's brief can broadly be described as: four storeys high, of brick construction, in two adjacent buildings, one of which is partially curved on plan. On face value this appears to be the most uneconomical solution that could have been sought, because the plan shape of the project and the number of storeys into which the accommodation is to be housed have equally as much effect on the cost of the completed building as the type of materials, or the form of construction, that have been used. Often, of course, it is the plan shape of the building that will determine the type of construction that is used. It should be realized from the outset that although this section of the book, at first glance, may seem distinct from the ideas contained in the preceding chapter, the points that are to be discussed in the following pages, should be foremost in the design team's mind, even in the early stages when the first ideas are being put on to paper.

If the Clapham Junction folly is not to be repeated, a knowledge is required about the plan shapes and storey heights, which, in a particular situation, will give best value for money. The enclosing ratio, which was used previously in the preparation of the cost plan, can now provide a means of comparison of the economies of various plan shapes. The enclosing ratio is a measure of how much enclosing wall, that is the enclosing envelope, inclusive of all doors, windows, etc., is required per square metre of gross floor area. Therefore, on the face of it, the lower the enclosing ratio, the comparatively more economical the design solution will be. For example, consider the three plan shapes illustrated in Figure 3, each single storey, five metres high, with the object of providing 100 square metres of floor space.

Although the gross floor areas are exactly the same for all three solutions, if the enclosing ratio is applied the following results will be produced:

Plan A $\frac{211}{100} = 2.11$ Plan B $\frac{261}{100} = 2.61$ Plan C $\frac{271}{100} = 2.71$

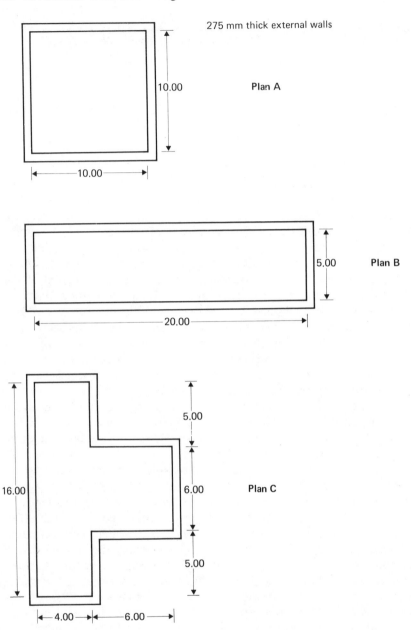

275 mm thick external walls

Plan A

Plan B

Plan C

Figure 3

In this case the first solution, Plan A, produced the best answer, in so far that, for every square metre of gross floor area there are 2.11 square metres of enclosing envelope required. It should also be noted from this exercise that the more the design progresses from the square plan shape, then the comparatively more expensive the design solution will become, when examined in this way. Indeed, the most economic solution expressed in this way would be a circle – a circle having the least amount of wall space enclosing the maximum floor area. Yet, as we all know, the construction costs for curved work are so high as to render this evaluation worthless. However, this is a greatly simplified view of the problem, because the shape of the proposed building may well be dictated by the nature of the proposed site, and finding the optimum design shape in a rigidly fixed set of co-ordinates is one aspect that will become increasingly more important as urban redevelopment is practised by local authorities throughout the country, and in these circumstances it is imperative that the design team is in possession of all the factors that contribute to economical design.

Plan shape will also affect the cost of the other elements of a building. It will have a direct relationship on the amount of internal divisions that will be required and whether these divisions need to be built, for example from load-bearing brickwork, or non-load-bearing demountable partitions.

The area of internal divisions, illustrated in Figure 4, that will be required in Plan A is 99.50 square metres, whereas in the case of Plan B it is 75 square metres, and although, originally, the square plan shape produces the most favourable results, the following factors now have to be set against these. Firstly, the spans involved in Plan B may allow the internal divisions to be non-load-bearing, as the roof structure could span across the shortest distance, and allow the partitions to be built directly off the oversite slab, resulting in a saving, because foundations for the internal partitions will not be required. If the square plan shape was adopted, then it is probable that at

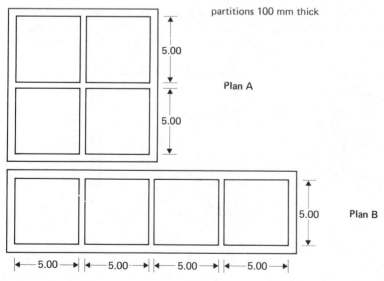

Figure 4

least two of the internal divisions would have to be load-bearing and consequently would require a foundation system. Therefore, not only will Plan B require less area of internal partitions, but also will implement a reduction in the cost of the type of materials that may be used, because the partitions and foundations may be dispensed with altogether. In addition to these items, there are other factors that have to be considered, for example, the larger the spans of the floors and roofs, then the proportionately more substantial and more expensive the construction will become, and this in its turn would probably affect the size of foundations. In short, the effects of a particular plan shape can be so multitudinous, that to discover fully the effects of differing design solutions a full cost study should be undertaken, and this will enable the various *pros* and *cons* of each design to be fully appreciated. In practice this operation is too expensive to carry out on every project, and more often than not it is merely done as an academic exercise. However, there is considerable benefit to be gained by the design team having a set of cost studies and comparative costs for standard sets of details at its disposal.

The choice may sometimes exist whether to provide the accommodation in one large building or two separate ones, the following exercise will make the decision easier. It is still assumed that the project is to accommodate 100 square metres of gross floor area.

Block 1 will require 131 square metres of enclosing walls and Block 2, 171 square metres, giving a total enclosing ratio of $\frac{302}{100} = 3.02$. In the problem that is being examined this solution is found to produce the most uneconomical plan shape to date. Hence, where design solutions are being considered, it is better to accommodate the floor area in a single building rather than in two separate ones. Taking the decision to provide two buildings will also have a substantial effect on communication installations, and electrical installations,

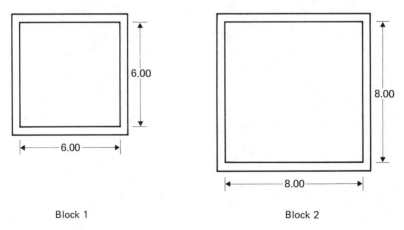

Block 1   Block 2

Figure 5

etc. From the heating engineer's viewpoint, the greater the area of external wall, then generally the greater the total heat loss and consequently the larger and the more expensive the heating installation will have to be, not only with regard to the initial costs, but also with regard to the subsequent running costs.

Until this stage the examples have been confined to single storey structures with a common storey height, that is, five metres. The architect may choose to provide the accommodation in a multi-storey building and the effects of storey height on the cost of a building will have to be considered. Depending to what use the building is to be put, the single storey structure is a comparatively expensive method of providing accommodation.

This is because most often, on normal site conditions, the size of the foundation system that is required to uphold the superstructure of a two or even a three storey building needs to vary very little in size and, consequently, cost from those required for a single storey building. It would seem then, that unless the client specifically requires accommodation to be in the form of a single storey building, instances of this may be found when the client asks for large uninterrupted floor areas for a heavy industrial process, it is not justifiable to recommend the use of single storey buildings.

If it had been decided to provide the 100 square metres of floor area in a two storey building, then the results shown in Figure 6 may be expected.

When the drawings are produced for this solution, it is discovered that the internal dimensions of the two floors have to be 7.07×7.07, but the overall height of the building need only be increased by 1.10 over the single storey building to 6.10 metres. The enclosing ratio, therefore, will be $\frac{186}{100} = 1.86$.

This solution, so far, seems to have provided the best answer as the enclosing ratio is considerably less than any of the previously considered plan shapes. As mentioned earlier, although the idea of providing two or three storey accommodation will, to a large extent, depend on the purpose for

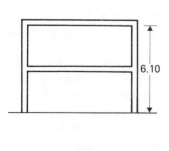

Section

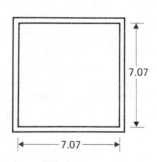

Typical plan

Figure 6

which the building has been designed, and in some instances it may be possible to provide three storey accommodation on the same size foundations that are required for a single storey building. In the case of structures more than one storey high it will be necessary to provide access to the upper floors, that is, a staircase or a lift, and areas of the gross floor area will have to be set aside for stairwells and/or lift wells.

Multi-storey buildings also have problems related to design and generally it is only the tremendous economies, that can be derived from the saving of building land by building upwards, that make this form of development a practical proposition at all. So scarce has building land become during the last few years, especially in the large cities, that design teams have had to think in terms of not only high rise buildings but also of sinking three or four basements, in order to provide the necessary accommodation, as has been recently done in the new building for King's College, University of London.

Not the least of the problems associated with high rise buildings is their indiscriminate location. The rash of tall structures that have appeared during the course of the last few years has resulted in the case of at least one London Borough, to the authors' knowledge, where the prevailing winds against the elevation of a high rise block caused eddies and mini whirlwinds at the base of the structure, making walking at an adjacent shopping precinct virtually impossible. So numerous were the complaints from local residents that eventually the shopping complex had to be roofed over.

In addition to these design and location problems the following factors contribute to the comparatively high costs associated with high rise blocks.

1   Generally, the taller the building the larger the foundations have to be in order to support it. Despite what was said earlier about buildings not only growing taller but also sinking into the ground, the provision of basement floors does not act like a flag pole in securing the structure in the ground and it is still necessary to provide substantial and expensive foundations.
2   When multi-storey structures were mentioned earlier, it was stated that allowances have to be made for access to the upper floors. Clearly the more storeys in the structure the greater the allowances that will have to be made, and these will contribute proportionately more to the overall cost of the building. In fact in the case of exceptionally tall structures the lift installation may have to be provided in two units.
3   Maintenance costs will be greater on a high rise building, a factor that must be taken into consideration when the materials for the façade are chosen. Provision for a permanent travelling cradle, suspended from eaves level, enabling all the areas of the external walls to be easily reached, may be necessary. The inclusion of such an item as this is becoming more commonplace when untried materials are used, and the amount of maintenance required is unknown at the start of the project.
4   The degree of fire resistance that will be required for buildings of this nature will have to be very high and, coupled with this, the means of

escape in the case of emergencies will have to be adequate. Both these factors increase costs considerably.

5   The actual cost of constructing a building at great heights is one that cannot be ignored. The *National Working Rule Agreement* allows extra payments to be made to operatives when they are working at heights. In addition, all the materials, plant and labour have to be hoisted to the area where construction is to take place.

Finally, a point that cannot be overlooked, as it has been so much in the past, is the effect that living and working on the thirty-fourth floor of a multi-storey building has upon people. In some cases it has been found that living at this height has had severe psychological effects on the occupants, that mothers live in constant fear that their children will tumble from the windows while playing, and in one instance it was discovered by a number of parents, that when their children were told that they were to be taken on a shopping expedition they proceeded to scamper around the flat, similar to a dog when its lead is shaken to tell it that it was time for a walk. Some old people have an inherent fear and distrust of lifts, illogical perhaps, but something that should not be ignored when the design of the building is first considered.

It would appear, then, that the members of the design team have to possess a very analytical mind in order to assess at the outset of a project all of the different factors that will affect the cost of the finished building. To help the design team sift through all the information and decide which parts are relevant, and which parts are not, in any particular case, it is possible to use the services of a computer. The application of the computer by the design team is mainly in the following fields.

1   As a library of information. This is perhaps the most basic advantage of using a computer, in that it is possible to store large amounts of information, infinitely more than the most conscientious design teams, and then at a moment's notice recall any part of it for use.

2   As far as basic design is concerned, as will be discussed later, the computer can help the design team to arrive at a well balanced decision, but to date very little use has been made of the computer's ability to consider all the facts and produce a set of working drawings.

3   It is possible to feed all the information that has been collected by the design team into the computer, which will then analyse it, and with reference to a set of standard details, with which it has already been programmed, and reproduce on a cathode-ray tube the standard detail that best suits the information with which it has been fed.

4   The computer can be programmed to perform standard calculations, for example, the enclosing ratios, or the areas of windows to gross floor areas. The relevant information has simply to be introduced into the machine and seconds later the results are available. This kind of service is not now only confined to expensive computers. The calculations that are included in this book have been checked by a relatively inexpensive

calculator, that can be programmed to perform such calculations by feeding into it a prepunched card, which could tell the machine how to perform the calculation necessary for production of the enclosing ratio. The actual figures can then be fed into the machine from a keyboard and the results appear almost instantaneously.

5   Throughout this book emphasis has been placed on good communications between the members of the design team, and a computer can be put to very good use in producing programmes for design team operations. Whether or not a computer is used on a project will, in the main, depend upon the nature of the design team and the costs that are involved in the hire of the computer.

In an attempt to obtain the maximum economies from any particular plan shape or storey height many of the building components that are available on the market are being designed to comply with certain standard sets of dimensions. British Standard 4011: 1966 *Recommendations for the Co-ordination of Dimensions in Building – Co-ordinating sizes for Building Components and Assemblies* describes the objectives of dimensional co-ordination in the following way:

Both the rationalization of traditional building and the introduction of methods of industrialized building involve the increased use of components fabricated on or off the building site. The co-ordination of the dimensions of components and those of the building incorporating them is essential in order:

1   To obtain maximum economy in the production of components, by variety reduction and by reducing the demand for special sizes.
2   To size components in order to achieve maximum flexibility in assembly, and so minimize the cutting of components and other site labours.
3   To increase the effective choice of components for building designers, by the promotion of interchangeability.

In order to achieve the co-ordination of dimensions it is necessary first to reduce the number of possible sizes that are to be used . . . .

The first selection of basic sizes, as defined in British Standard 2900: 1970, for the co-ordinating dimensions of components and assemblies should be, in order of preference, as follows (when n is any natural number including unity):

First: $n \times 3$ millimetres
Second: $n \times 1$ millimetre
Third: $n \times 0.5$ millimetre up to 3 millimetres
Fourth: $n \times 0.25$ millimetre up to 3 millimetres

The third and fourth preferences should not be used for basic sizes over 3 millimetres unless there are strong economic or functional reasons for doing so. The fourth preference is put forward provisionally. There may be need for other sizes below 0.5 millimetre, but there is as yet insufficient evidence on which to base a firm recommendation.

Standard Steel
WINDOWS AND DOORS

CHART 1

Shaded spaces will be filled with single or multipane standard window units. Details of types of casement and of work sizes will be published by individual manufacturers. Flexibility of the basic matrix is increased as follows :

Length

1 by including one or more fixed light units 500 mm long in the composite window

2 by including one or more pressed steel box mullions or partition covers in the assembly. Available in all matrix heights, these add 100 mm to length.

3 by applying a wood surround, adding 100 mm to length and height.

Height

1 by combinations of basic units to fill the intermediate heights,

e.g. $\underline{500} : \underline{500} : \underline{500} : \underline{500} : \underline{500} : \underline{200} :$
$\quad 500 . \quad 700 \quad 900 \quad 1100 \quad 1300 \quad 1500$

2 by applying a wood surround, adding 100 mm to height.

Thus all modular lengths and heights from 900 mm upwards can be achieved in increments of 100 mm.

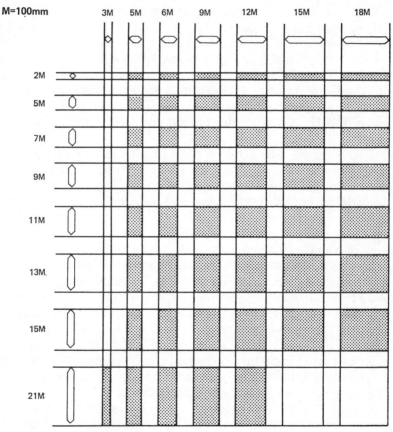

*Now Referred to as 'Co-ordinating Sizes'

Figure 7

Of the many organizations that co-operated with the British Standards Institution in the preparation of the recommendations contained in BS 4011 and BS 2900, the Steel Window Association in collaboration with other similar organizations, in both the public and the private sectors, agreed upon metric modular co-ordinating sizes for steel windows and doors, for use with a modular grid. Very briefly, a modular grid can be described as a three-dimensional grid into which the proposed building is designed. The grid does not dictate the size of the steel window or door that may be used, but allocates a basic space, into which it may be fitted. Allowances must be made, when designing, for wooden surrounds, jointing, and so on, and this system of design is not as limiting as it may first appear. For example, a 75 × 75 mm wood surround, would add an extra 100 mm to the size of the window, with allowances for rebates and grooves, etc. This is perfectly acceptable as it conforms to the second preference dimension in BS 4011.

Figure 7 shows a chart, taken from one of the Steel Window Association's publications and reproduced here with their kind permission, illustrating the co-ordinating sizes selected for standard and steel windows and doors. The lengths of the co-ordinating sizes conform to BS 4011 first preference in increments of 300 mm. Shaded spaces will be filled with single or multipane standard window units. In addition to this all modular spaces conforming to BS 4011 second preference in increments of 100 mm from 900 mm upwards can be filled by a combination of units and provision is also made for purpose made units.

Clearly then, when the time comes that all building components are designed and manufactured to a well-thought-out set of recommended dimensions, like those laid down in BS 4011 (and the change-over from Imperial to SI units for measurement, provides an excellent opportunity to achieve this), it will be possible to obtain maximum economy in the production of components and also to move nearer to maximum efficiency on site when the components are incorporated into the structure.

It is just a short step from the dimensional co-ordination of building components to industrialized building. There are now so many systems of industrialized building available on the market that it is not really possible in a book of this kind to say categorically that industrialized building will or will not produce a more economical result to a design problem. The principal reason for this is that many systems of industrialized building derive their greatest saving from the fact that many of the components that are used are manufactured to standard sets of dimensions, in a kind of production line atmosphere to be assembled quickly on site. To achieve the maximum economics from this type of system it is generally necessary to keep the production line running as near to continuous output as possible and unfortunately this is not always possible.

It is a fitting conclusion to this chapter to mention that within 500 metres of the building described at the beginning of this chapter a vast complex of high rise flats has been erected, using industrialized building techniques.

3 Design criteria

Builders, property owners and members of the building design team are becoming increasingly aware of the relationship that exists between the design of a building and its eventual overall cost. Consequently increasing pressure is being exerted on the design team to evaluate, before a project leaves the drawing board, the suitability from the cost and performance aspects of the design and the forms of construction and materials that it is intended to incorporate into a proposed project. Over the years it has been proved that the selection of the correct design, coupled with the correct choice of materials and systems, have long-term effects on the performance of a building. This evaluation usually involves an examination by the quantity surveyor of all the viable design alternatives at various stages in the design sequence. To do this it is necessary to know the expected performance and the overall costs of all the major components within a proposed project. The type of information that should be available before the evaluation of a particular component or element can begin is, for example:

1 The design brief and user requirements
2 The initial and long-term costs associated with a component or element
3 What maintenance costs, if any, are likely to be incurred and at what intervals
4 The expected life of a component or element
5 In the case of a component, the expected life of the structure into which it is to be incorporated. (It would be an unnecessary expense to build into a structure with an expected or planned short life, materials of a highly durable and maintenance-free nature.)

If short life buildings were cheap enough, a series of them might cost no more than the equivalent long life building and would appear to have the advantage that the design brief, that is the user requirements, might be revised every few years to keep the building up to date and the design working effectively. A report by a committee formed by the Department of the Environment found that there may be some economic justification for the use, not of short life buildings, but of short life finishes with a maintenance free life. These could be incorporated into a more durable structure allowing the internal arrangement, or the finishes of the building, to be altered during the expected life of the complete building. This principle is already applied to the design of shop fronts in fashionable town centres, where the façade must change almost as regularly as fashions do. With proposals like these it is not too difficult to see a time – indeed that time is almost upon us – when the design of a building will be like completing a giant jig-saw puzzle; the pieces of the puzzle will be building components, all with different performances, life spans and maintenance problems. This prospect makes the increased use of design evaluation even more necessary in order to use compatible materials successfully within the same structure.

The money and manpower resources spent annually on maintenance are a substantial proportion of the country's total building activity. Furthermore, for maintenance, the output per man in money terms is much less than the equivalent for new work, with little chance of much increase in this output owing to the limited size and nature of maintenance work.

Present day interest rates, tax laws and shortage of capital often indicate that it is more economic to avoid the more expensive materials which require low maintenance and to select instead low cost components requiring high maintenance; but this indication takes no account of the inconvenience, disruption and loss of earning potential which arises with the need to repair or replace. There is no way of measuring this factor in national terms, or indeed in many individual cases, so it is perhaps frequently ignored.

With the stock of buildings increasing in number year by year, the problem of future maintenance is continually growing, and the situation will become worse as the available labour finds more rewarding employment elsewhere.

The conflicting demands on the available capital resources to the nation or to any individual authority inevitably lead to the pruning of budgets. The result is that building cost limits are usually set at a level which leaves very little scope to allow a serious attempt to be made to reduce the problem which is being created for the future. There is a similar reluctance in the private sector to invest in higher initial costs in order to make savings in the future.

Another fact to consider is the uncertainty of the future, with changes due to company take-overs and mergers, and individuals moving more frequently from house to house, it is clear the unnecessary expenditure should be avoided unless it can be reclaimed in an increased value on sale.

A source of reference for such data as are required to evaluate a building design is the Building Maintenance Cost Information Services. This is a system for collecting and interpreting building maintenance and other property occupancy costs including cleaning and maintenance with relation to the expected life of particular components. The information is in a form that can be easily and meaningfully used by the building economist when comparing differing design solutions.

The importance of using materials and a form of construction that has been carefully evaluated can be illustrated by means of the following example. Since the war an alarming number of multi-storey structures have been constructed, both by local authorities and by private developers. At first one of the more popular, and it was thought less complicated forms of construction, was a reinforced concrete framed structure, with brickwork infill panels. On the face of it, a well-tried form of construction which offered no problems with either erection or maintenance. However, during the last few years numerous examples of severe cracking of the brickwork cladding have been reported. Cases have arisen in Leeds, one of the pioneers of high rise flats, Plymouth, Salford, Hull, Lewisham (London) and Clydebank. This means that the clients involved have now to pay considerably more than was originally envisaged for maintenance over the life of the building.

Many of the design characteristics of buildings are a direct outcome of design decision, or the quality of the construction. Part of the knowledge and skill demanded of the design team is the ability to take into account the continuing technical and economic consequences of design decisions and be able to evaluate the future operating costs of buildings.

When comparing the costs and the performance of various building design solutions, not only the initial costs are taken into account, but also where applicable any recurring costs due to maintenance, partial replacement, etc., that may reasonably be expected throughout the life of a building. Taking into account the effects of the costs of maintenance and so on may seem inconsequential, until it is realized that in 1969 the cost, in this country, of building maintenance was in excess of £2000 million.

Consideration, when designing, should be given to the way in which the completed building is to be disposed of. If the client is a developer the finished building will either be:

1   Sold or let for profit
2   Used by the client for his own use

In the first instance the initial costs of any services, etc. will be the responsibility of the client and will have to be paid for out of capital, which in turn will probably be borrowed, perhaps at a high rate of interest. If the completed building is sold or let then any subsequent running and maintenance costs will be paid for by the occupier and not the client. Therefore, in this case, finishes and construction systems with low initial costs but high subsequent costs will be most beneficial to the client.

If, on the other hand, the client intends to use the building himself then obviously the maintenance costs will have to be paid for by him. Initial costs come out of capital and therefore are not eligible to be used as an allowance against income tax; however, maintenance costs can be used in this way and, therefore, are of benefit to the client in this respect.

In order to evaluate a design solution from the cost and performance aspects it is necessary to carry out a study of the constituent parts or elements of the building under construction. Although these studies are carried out on an elemental basis and therefore each part of the building tends to be thought of in isolation, it should be remembered that many of the elements are functionally independent and the decision to change the construction of part of a building, for something apparently more beneficial, may in turn have an injurious effect on some other elements.

For example, it may be found that to substitute a cheaper form of material in the external walls of a building results in an increase in the heat loss, which in turn means that a larger and more powerful form of heating system will be required to maintain heat levels.

Almost certainly a larger system will cost more than the original system, both initially and in the long term, and may eventually outweigh any economic advantages that were thought to have been gained.

The economics associated with the design of major elements will now be considered, a major element being defined as a part of a building which generally contributes a high percentage to the total cost.

Substructure

The bearing capacity of the ground has been described as 'infinitely variable' and this comment can equally well apply to the design or the choice of foundation systems. It is this variability and uncertainty that makes this element so comparatively difficult to evaluate. There can be said to be four basic types of foundation systems, these are:

1 Strip foundations
2 Proprietary systems of foundations
3 Raft foundations
4 Piled foundations not included in (2)

The type of foundation system that will be used for a particular building will, to a large extent, be dictated by the physical characteristics of that building. Research and experience have proved that traditional strip foundations can be safely and economically used for buildings of up to four or even five storeys. At the other end of the scale the concept of multi-storey buildings would to a large extent be impossible if it were not for the use and the increased development of piled foundations, in all their forms. However, this is an over-generalization of the situation and there are many occasions when the relative economies of various types of foundation systems have to be examined, especially in the cases of low and medium rise buildings.

The type of foundation system that is probably used most often throughout the country is the traditional strip foundation. This system has the advantage of being easy to construct, generally requiring no special equipment; however it does suffer from the distinct disadvantage that it is labour intensive. An analysis of the costs that are involved with this element shows that probably more than any other element the majority of the total costs are attributable to expenses of labour. With the ever-increasing cost of employing labour in all its forms it is only logical that attempts should be made to reduce the percentage of labour costs within the total costs of the element. Figure 8 shows how labour costs have risen during recent years, compared to the increase that has taken place generally in the cost of materials. There is every indication that this trend will continue.

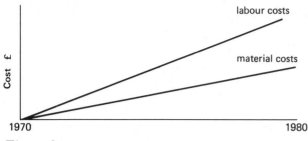

Figure 8

Traditional strip foundations are used mainly for low and medium rise buildings, where the bearing capacity of the ground presents no problems and where there are unlikely to be any heavy point loadings. However, the rapid increase in the cost of labour has made the use of these foundations, which were for a long time considered to be an inexpensive design solution, an uneconomical proposition.

If present trends continue and labour remains such an expensive commodity then the relationship that labour costs bear to the total cost, not just of this element, but of the whole building process, is a very significant factor in determining the price of buildings. As inevitable as the reduction in the amount of man hours for this element sounds, in order to achieve this it is necessary to introduce some form of industrialization and/or mechanization into the building process. In the early 1960s when various forms of industrialized building were being investigated, The Basingstoke Development Group tried to produce a system of industrialized building for two storey houses. As with most systems of this kind it was found to be a relatively easy task to 'industrialize' the elements and the components of the work above ground level. However, when it came to the substructure it was found to be impossible to 'industrialize' any of the operations, leaving a total of 166 items that had to be performed either totally by hand or with the aid of mechanical plant. This is still true today.

Of the attempts that have been made to reduce the number of operations involving labour with this element, two noteworthy attempts are the use of deep strip foundations, even on sites where the bearing capacity of the ground is adequate, and the so-called *Finchampstead Project*, which investigated the use of factory-made components in the substructure. Both of these systems fall into the category of proprietary techniques and would therefore seem to suffer from serious financial set-backs which usually, in the final reckoning, make these inherently cheaper forms of construction only break even as far as costs are concerned. It appears that, especially in the case of the *Finchampstead Project*, the maximum economies are only to be obtained when it is possible to produce the components on which this system is based at a constant output over a reasonable period of time and when builders have become as familiar with new and developing techniques as with traditional forms of construction.

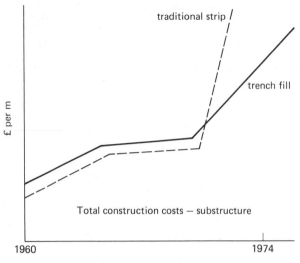

Figure 9

Firstly, the use of deep strip foundations. The Cement and Concrete Association have, in the light of rising labour costs, promoted a system called 'trench fill' for constructing foundation systems for domestic dwellings. Figure 9 shows the comparative costs of trench fill compared to traditional strip foundations. It seems that the use of this type of foundation, which is material intensive, is an economic proposition in certain circumstances.

Reducing the amount of labour intensive operations required to construct a foundation system, in a manner described by the association, is claimed to produce savings of up to 33 per cent compared with traditional strip foundations. A detailed comparison was carried out by S. Lazarus and Partners in association with Peter H. Hill and Partners and the subsequent report shows that on a typical block of five three storey houses there can be a saving equal to approximately 6 per cent, and on a typical block of two storey detached houses there was a total saving of approximately 31 per cent. In times of galloping inflation these figures at first glance appear to be outstanding; however, savings are cut if work cannot be carried out under near perfect conditions, that is on a dry, level site. But, conditions apart, this system has distinct advantages over labour intensive methods and is worthy of development.

The importance of developing an economic system of foundations for low and medium rise buildings lies in the fact that, as the latest housing and construction statistics show, only very few homes were approved to be built in high rise buildings. The construction of dwelling houses, in both public and private sectors, including associated works, account for approximately 35 per cent of the total turn-over in the building industry.

The *Finchampstead Project* was a collaborative exercise between the

Department of the Environment and the Building Research Station. A group was appointed to design a housing scheme for a site of five hectares in Finchampstead and it was decided that all the houses were to be terraced with a rectangular plan shape. The site was flat and presented no problems. Various methods were considered for providing a foundation system for the houses and despite the apparent simplicity of raft foundations the man hour requirements were high, about one third of the man hours required subsequently for the superstructure as a whole. Therefore, the group set to work to design a more economical system which briefly involved casting *in situ* concrete foundation pads in a foundation trench, levelling these pads and then laying factory manufactured pre-cast concrete beams between the pads. It was possible to arrive at four beam lengths that in various combinations could make up all the lengths and depths required. The successful tender figure was slightly higher than the sum estimated for traditional foundations, but this, as has been stated earlier, may be due to the contractor's unfamiliarity with the method of construction. However, on site there were practically no subsequent complications and the final overall cost indicated a saving of 10 per cent on the estimated cost of traditional strip foundations.

The effects of the cost of maintenance of foundations is seldom, if ever, a deciding factor in the choice of a particular system. This is due to the fact that a foundation's function is so basic that if designed and constructed properly, it should function adequately throughout the expected life of the building. Even with the comparatively untried methods as described above, it is doubtful whether there would be any maintenance costs to be taken into account and in any event to design a foundation system with a limited life span would indeed be a risky operation.

Raft foundations have rather special application and are used generally where no firm bearing strata exist at a reasonable depth and where loading is reasonably light and evenly distributed. This form of foundation may be the way to carry the load from lightly loaded columns to weak soil, but it suffers from the disadvantages associated with strip foundations, that is, it is labour intensive. In fact, during the investigations to choose a type of foundation system for the *Finchampstead Project* it was discovered that although raft foundations may have provided a reasonably simple method of construction, about half the man hours required for the substructure as a whole were simply for laying services on, or under, the ground floor raft.

It is sometimes thought that piled foundations are necessarily an expensive solution to a foundation design problem and of course, on the face of it, a direct comparison of costs with most other foundation systems will show this to be correct (see Figure 10). However, whenever piling in one of its many forms is used, it should offer an economically better solution than the other forms of foundation systems that have previously been discussed. Piled foundations are used principally in the construction of high rise buildings and the savings that can be achieved by building upwards instead of outwards and the total floor area it is possible to provide on one set of foundations, over

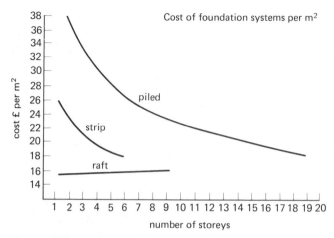

Figure 10

which the cost is spread, making piling an economic device. One considerable advantage piled foundations have over other forms is the fact that construction need not be held up by periods of rainy or otherwise inclement weather.

Finally, the plan shape of the proposed building will have an important effect on the cost of this element, not only in the ways described in Chapter 2, but if the plan shape of a building develops from a square then the site costs involved in setting out the substructure also increase considerably. Recently a number of private developers and local authorities used multi-sided structures in an attempt to achieve the optimum design solution based on the facts that:

1   The external walls of a circular plan shape will enclose the greatest amount of floor area of any plan shape but are expensive to construct in practice when working with curves
2   The square is the most economical plan shape to use in practice

Although exact figures are not available, the London Borough of Hillingdon have achieved a degree of success and produced an economic solution with a development of multi-sided shop units within the borough.

Therefore, the cost and design criteria to be taken into account when considering this element are:

1   The physical characteristics of the superstructure, paying particular attention to the following elements: frame, upper floors and external walls, remembering that no element should be considered in isolation
2   During present economic trends endeavour to minimize the amount of labour that is required
3   Consider the project to establish whether it is large enough or repetitive enough to benefit from the use of industrialized building techniques, for example, a housing scheme

4 Consider the best plan shape and the nature of the site
5 Finally, try to establish if there are likely to be any maintenance costs during the expected life of the element

The next major elements to be considered are:

Frame/upper floors

Frame

Generally speaking, it is not necessary to use a frame for buildings with normal loadings up to four or even five storeys high, as it is possible to use load-bearing walls economically, together with traditional strip foundations described earlier in this chapter. However, for taller buildings, where the total of the dead and superimposed loads becomes considerable, then it is usually not economical to use load-bearing external walls. They have to be so thick to withstand the loading that the usable floor area can be cut down considerably, and therefore it is better to use some kind of framework. The object of a frame is to transmit the loadings, via a series of beams and columns, to the foundations, thereby allowing the external walls to be comparatively thin.

This statement is a sweeping generalization and there are, as with most things, exceptions to the rule, for example the World Trade Centre, New York, dubbed when it first opened 'the tallest building in the world'. It provides 10 million square feet of rentable office space in two, one hundred and ten storey, 1350 feet tower blocks. The blocks were designed so that 75 per cent of the plan area is rentable space and yet this has been achieved by using a system of load-bearing external walls which completely eliminates the need for an internal system of columns. At the other end of the scale, steel framed two storey houses are being constructed on quite a large scale in the Midlands and the North of England. The system, its manufacturers contend, gives 'unlimited choice of design possibilities' and is economically viable due to ease and speed of erection on site.

The most commonly used types of frame are:

1 *In situ* concrete
2 Pre-cast concrete
3 Structural steelwork

The case for concrete against steelwork

The regulations relating to building in this country stipulate that buildings intended to be used as hotels, offices, etc. should have good resistance to fire. Concrete has an inherent resistance to fire, whereas steelwork will buckle and twist when exposed to intense heat. Ironically it is often protected from fire damage by being coated with its own best competitor, concrete, although of course there are a score of other materials that are widely used for this purpose.

Steelwork therefore has the major advantage in that it is speedy to erect, much of the fabrication can be carried out in 'shop conditions' and is therefore unaffected by periods of inclement weather. In addition the initial cost of steelwork is less than its rivals, however, much of this advantage is removed when the cost of the necessary fire protection is added.

But not all buildings require such a high degree of fire resistance and in these areas steelwork has major cost advantages over its competitors. For example, it is ideally suited for use in single storey warehouses or factories, together with lightweight cladding to the external walls.

In a project where speed is of the essence, as a general rule, pre-cast concrete structural members are ideal. However, they are expensive and if it is decided to use them, the size of the members should be kept to a minimum. Also the site should not be so restricted as to hinder the delivery, unloading and transportation of, what could be, large pre-cast beams.

Finally, timber is coming more on to the scene as a structural material, its greatest disadvantage being its propensity to burn. However, recently the Bell Sports Centre in Perth was constructed using a laminated timber beam dome providing 3000 m² of floor area. Shortly before its opening it was engulfed for an hour by fire, even so the structure suffered little damage.

Timber is also an economical form of construction to use for the first floors of domestic dwellings. In fact, timber-framed bungalows for use with extendable homes are becoming an economic solution to the problem of an expanding family. For larger and more complicated structures a number of reinforced concrete ribbed, trough and waffle floor sections have been developed with the object of reducing the volume of concrete required and therefore the cost. Pre-cast plate floors are available and are much more speedy to erect than the *in situ* systems described above. However, as we have seen before, pre-cast members are more expensive than the *in situ* ones.

Therefore, because each form of construction, used under ideal circum-stances, offers saving in time and/or overall cost, the problem is deciding when to use what system and where, for this reason, it is best to evaluate each set of circumstances as they occur.

However, the size and arrangement of the columns can also be a critical determining factor when deciding on the type of frame. If the client is building for letting the efficiency of the design and especially the ratio of the gross to net lettable floor area is of utmost importance, because it is undesirable to have large areas of floor space taken up by columns. For example, in the City of London where office rents are very high, a tenant may argue that the areas taken up by columns should be deducted from the lettable floor area, hence a loss in revenue. In such a case it is more beneficial to have a few carefully positioned columns.

Like the previous element, substructure, there are a very few long-term maintenance costs associated with the frame or upper floors of a building, with the possible exception of repainting the exposed steelwork of a factory or warehouse every five years or so and therefore the initial cost of this element is likely to be the only cost. That is unless some sophisticated system like the one used for the new London Stock Exchange is specified, for it has to be estimated that it will cost as much to demolish the pre-stressed structure in eighty years' time, as it has to construct it, because of its complex 'umbrella' type structure.

Therefore, the cost and design criteria relating to this element are likely to be:

1   The functional requirements of the building
2   The type of external walling or cladding to be used
3   The order of priorities: for speed choose pre-cast, for economy *in situ* concrete
4   Does the structure lend itself to the use of a particular material, for example long, high sweeping concrete vaults and spans
5   Finally, the comments relating to labour efficiency and plan shape that were made in the element substructure, still apply here

Roofs

The RICS Standard Form of Cost Analysis requires that the cost of the roof element should be split into the following:

1   Structure
2   Coverings
3   Drainage

After this it will be found that the first two parts, structure and coverings, will make up the majority of the elemental cost. The form of roof and its coverings will, to a great extent, be dictated by the use to which the building is to be put. For this reason, we have attempted the dangerous practice of generally categorizing building types thus:

Commercial and domestic
In this category the choice is often between pitched or flat roof construction. A cost comparison between a timber-pitched roof and a concrete flat roof, excluding the coverings, will show a concrete flat almost 35 per cent more expensive than a pitched roof. However, savings as great as this may only be achieved with regular plan shapes: small insets or projections or other complications in the plan shape of a building with a timber-pitched roof will considerably increase the number of hips, valleys and intersections which will in turn increase the material, the labour content and the cost. Therefore, once again, the effect of the plan shape can be seen to be an important influence on costs.

The rule seems to be, for high rise buildings to have flat and not pitched roofs despite the savings outlined above. The main reasons are that pitched roofs on tall buildings look so incongruous and also that the roof area on tall buildings is used for plant rooms, tank rooms, etc.

Industrial

Industrial buildings generally need roofs that provide large uninterrupted spans and also are required to incorporate some form of roof lights. The cheapest solution in cases like this has been found to be an asbestos cement sandwich with translucent sheets on lightweight mild steel roof trusses. For buildings requiring exceptionally large spans or a higher degree of fire resistance there are a multitude of proprietary systems and materials some of which are capable of being erected very quickly, but at a higher cost than the system described above.

For flat roofs the choice of coverings seems to be sheet metal, asphalt or felt. Metal sheet roofing needs no regular treatment other than the removal of accumulated rubbish, and an inspection of joints and areas where the metals are stressed. Inspection of asphalt is advisable at intervals. The material is brittle when cold, and mechanical damage can occur. There may be blistering where there is no underfelt (for example, on steep slopes or up-standards), or signs of crack formation caused by structural movement, or general minor shrinkage causing movement around edges. All bitumen felt roofs should be inspected on a routine basis.

Regular treatment of bitumen felt is advisable only where the bitumen is exposed, but most such felts have mineral surfaces of some kind, which cannot be adequately treated, though minor local defects can be repaired by patching. Anything other than minor defects calls for replacement. Asphalt roofing properly designed and laid should prove capable of lasting 50–60 years; the natural ageing of bitumen felt is likely to limit its life to about 20 years.

Sheet metals may also, of course, be used for pitched roofs although this will prove to be expensive. The most economical and widely used type of covering for pitched roofs being concrete interlocking tiles on battens and felt. Metal sheet finishes are generally the most expensive, and minimum specification built up bitumen felts the cheapest. Asphalt may not be more expensive than the more sophisticated built up finishes or plastics. Natural rock asphalt is about one third more expensive than that with limestone aggregate. Costs vary with the complexity of the surfaces to be covered, and from this point of view, the design of roof surfaces should take careful account of the material to be used.

The durability of flat roof finishes is a major cause for concern, and it is advisable to consider at an early stage in design the recommendations of the relevant trade associations with regard to suitability for purpose within the cost target. Statements about the life expectations of roofing materials refer to perfect laying, which depends on the efficiency of site operations.

In calculations of cost, repairability is also important, particularly in the lower price ranges, where exceptionally long life is unlikely to be achieved. Fundamentally, only a well-built, well-specified roof is likely to prove economic in service.

What is the future for plastics in roofing?

The plastics industry now has probably the greatest opportunity in its history to make a significant impact on the roofing market. With the daily increasing world shortage of fuel, the need to conserve energy is becoming of paramount importance. This will inevitably lead to the introduction of higher standards for insulation in buildings, particularly in roofs. The use of the relatively inefficient traditional wet insulating materials is therefore likely to decline, in favour of compact, efficient, inherently dry insulants such as plastic foams. However, these materials will only be used successfully if the manufacturers and roof designers fully appreciate the actual environmental conditions which these materials will experience on site and also the misuse to which they will inevitably be subjected. Lack of understanding of these factors has probably made a major contribution to the disappointing performance of plastics materials in roof construction so far, particularly when used for waterproofing. This situation is likely to be exacerbated in the future in conventional roof designs since the roof coverings are likely to be subject to even more thermal stress than at present as the insulation standards are increased.

Roof designers and manufacturers will, in future, need to give much more thought to the concept of the total roof system and not just consider the individual components in isolation.

Therefore, the factors that influence the costs of this element are:

1   The plan shape
2   Whether or not a pitched roof is suitable
3   The type of coverings to be used
4   The life span of the building

Stairs

The cost of the stairs is unlikely, even in the tallest building, to be a major contributor to overall building costs, and their inclusion in a building can hardly be termed an extravagance.

However (and this seems to apply particularly to the finishes and balustrading), it does seem to be an element that falls into that unfortunate category of 'possible savings', and for that reason its cost may well come under close scrutiny. For analysis purposes it should be remembered that the costs can be divided into:

1   Structure
2   Finishings
3   Balustrades

1 internal stairs/service core

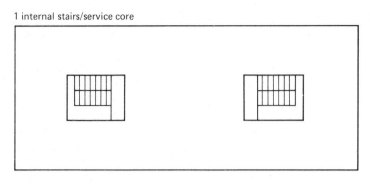

2 external stairs/service core

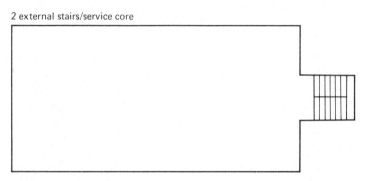

Figure 11

Also when allocating costs it should be borne in mind that for very tall buildings, or even low rise buildings that are to be crowded with people, stairs and stairwells have to provide a high degree of fire resistance.

Although the actual cost of stairs is not usually a critical factor, the positioning of stairs, particularly in high rise blocks, can be. Figure 11 shows two basic alternatives:

1   With the stairwell/service core integrated into the curtilage of the building
2   With the stairwell/service core positioned externally

Alternative (2) will generally be the most economical solution but also the most likely to be rejected aesthetically. (2) is more economical because it produces a better overall gross/net floor ratio, meaning that the overall floor area in (2) can be less than (1) and still fulfil the client's brief, with less area devoted to unprofitable circulation space.

External walls

This is the element in which the architect can impress his personality on the new building, it will determine how the world will view his design and, therefore, quite naturally is of great importance to him. It is also an element which contributes a great deal to the overall cost of a building, and one where large areas are involved and an increase of £4 per square metre may sound insignificant until multiplied by 4000 m² of external walls.

As explained previously the plan shape will determine the wall/floor (or enclosing) ratio, and consequently, the cost of the external walls and windows elements. Table 2 shows how, for an average office block of reinforced concrete frame structure, the construction costs may be expected to be distributed. The plan shape will have an effect on all the largest contributors to cost and especially on the external walls. Therefore, the enclosing ratio is of great importance in the design concept of a building, remembering that the lower the ratio, the better the value for money, and in a well designed office block a wall to floor ratio as low as 0.5:1 to 0.6:1 can be achieved (see Figure 12).

Preliminaries and contingencies	16%
Substructure	6%
Superstructure	17%
External walls	16%
Services	34%
Finishes and fittings	9%
Drainage and external works	2%

Table 2

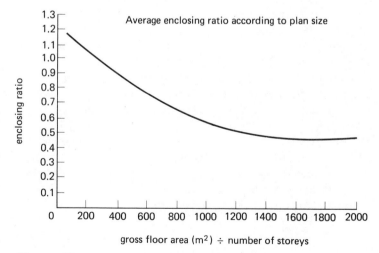

Figure 12

For low and medium rise structures load-bearing external walls are the most economic solution. Numerous attempts have been made over the last decade to industrialize and prefabricate external wall panels for domestic dwellings. The biggest advantage of prefabrication being, of course, speed. However, traditional brickwork has recently 'hit back' and a pair of traditionally built brickwork semi-detached houses were, according to the builders, completed within eight working days.

For high rise framed buildings, that are to be used as offices or flats, the choice of infill panels and curtain walling systems is vast, the principal governing factor being cost. It is perhaps a symptom of the times in which we live that the availability of building materials and consequently their cost is cyclical. At the time of writing 'cow coloured' bronze tinted glass is easily available and reasonably inexpensive, with the consequence that a rash of speculative office blocks are appearing incorporating this material.

As the external walls will (irrespective of height or plan shape) envelop the building, their ability to transmit or retain energy is of great importance.

The cavities of cavity wall construction should be filled with insulating foam, and large areas of glass should be double-glazed. These provisions do, of course, add to initial costs, but in the long term will help save and keep down the increasing cost of energy. The decision will have to be taken whether or not to incorporate extra insulation into the structure and the outcome could depend upon the use to which the completed building is to be put. A client, building for owner occupation, will be more concerned about the size of the electricity and heating bills, than if the building is to be let to tenants, when the responsibility will rest with them.

There are so many types of external walling for multi-storey buildings and they have nearly all been introduced so comparatively recently that it has not yet been possible to evaluate the relationship between initial and long-term maintenance costs. For example, Figure 13 shows a recently constructed office block of a particularly distinctive design.

How the façade detail will stand the test of time, what the long-term maintenance costs will be, and what it will look like twenty years hence is a matter purely for speculation. Sophisticated systems of curtain walling and fenestration on large structures may contain literally thousands of linear metres of waterproof joints, that will undoubtedly require some form of maintenance, in addition to general cleaning. Therefore, in cases like these, provision should be made for a maintenance cradle at roof level, as the only practical method of carrying out running repairs.

However, using techniques that are fully described in Chapter 4, a costs-in-use exercise may be reliably carried out on the basis of the materials that have been used for some years now for the external walls of low and medium rise buildings, viz.:

1 Brickwork
2 Pre-cast concrete
3 Timber

Figure 13 Angel Court, Throgmorton Street, London EC2
Photograph courtesy of Ove Arup & Partners

similar to the exercises included in Chapter 4. Therefore, it may be said that the design criteria to be followed when examining the external wall element are:

(a) Endeavour to produce the most economical enclosing ratio possible
(b) Use materials that are in 'vogue' and hence easily available
(c) Use materials with established maintenance costs (but beware! see pages 98–9)
(d) Incorporate standard details at eaves, external angles, etc.

Windows

Windows tend to be taken for granted, yet they are an important element of the structure, providing light, ventilation and much of the character of the building.

The design of windows extends to double-glazing and double windows which assist considerably towards good thermal insulation and acoustic values.

In the following pages alternative costs of some of the available materials for the manufacture of windows are compared, not only in terms of initial expenditure but also in terms of cost over their life in the building, assuming that a regular pattern of cleaning, repair and redecoration is maintained.

There is a wide choice of materials from which windows can be manufactured, the most commonly used being:

Timber
It is becoming increasingly difficult today to obtain timber of the right quality that has been satisfactorily seasoned. Twisting or warping can lead to badly fitting opening casements with consequent draughts. Failure to protect the timber with adequate weather details can lead to attacks of wet rot requiring subsequent renewal.

Steel
Without adequate maintenance, steel is subject to rust, which gives rise to difficulty in control and to bad fitting.

Aluminium
Aluminium is a metal that does not require protection to prevent it from corroding. With a mill finish, it is subject to chemical action when exposed to the weather and this causes powdering of the surface and pitting of the metal. Frequent cleaning is therefore necessary.

Anodized finish
An anodized finish gives considerable resistance to atmospheric conditions and merely requires regular washing.

Plastic
Solid PVC and fibreglass windows are available, but as yet are comparatively recent innovations. Their qualities and performance therefore have not yet stood the test of time.

Stainless steel
This material is expensive although it should have a long maintenance-free life. A stainless steel finish can also be applied to aluminium sections.

External doors

The selection of material and form of construction for external doors is usually governed by cost considerations, and as with most commodities, the reliability is in a direct relation to the initial cost. In a building, this is not a satisfactory state of affairs, as breakdown of basic elements shortly after initial occupation is unacceptable. Where cost limits preclude the use of naturally durable materials for the external fabric of a building, whether doors, windows, or cladding, measures must be taken to ensure that acceptable life expectancy can be guaranteed. With external doors, the most economical assemblies undoubtedly consist of timber constructions of various forms, and it is these which require the most careful consideration at the time of design or selection to obviate failure caused by the inherent weakness of the material. Quality of timber, preservative treatment, protective systems and their subsequent maintenance are all essential to proper functioning for an acceptable period.

Structural failures are less likely where metal or glass components are used, but the architect must satisfy himself that the assembly will meet the necessary standards of performance and that the manufacturer's sales orientated literature does not conceal possible inadequacies in standards of finish and methods of fixing and support.

Internal walls and partitions

The vast choice and cost of internal walls and partition materials range from plastered blockwork to sophisticated demountable partitions, and can be supplied by the client as part of the 'fitting out' contract.

Internal doors

As is the case with internal walls and partitions this element does not contribute a large percentage to the total building costs, and listed below are the relative costs of internal door types indicated by a cost index. It has been assumed for the purpose of the index that doors are installed and include wall openings, frame, ironmongery and finishes.

Door type: flush
100   Unlipped, hardboard faced, cellular core, softwood frame, painted
105   Lipped two edges, sapele plywood faced, cellular core, factory fully finished with clear seal, painted softwood frame
109   Lipped two edges, plywood faced, cellular core, softwood frame, painted
118   Lipped two edges, plywood faced, semi-solid core, softwood frame, painted
136   Lipped two edges, plywood faced, solid core, softwood frame, painted
140   Lipped two edges, faced with laminated melamine in standard colours bonded to plywood, cellular core, painted softwood frame, 40 mm thick

200 Teak, lipped two edges, hardwood veneered plywood faced, solid core, fully finished, clear seal hardwood frame
127 Lipped two edges, plywood faced half hour fire check door, equivalent to BS 459, Part 3, painted
175 Lipped two edges, plywood faced, one hour fire check door, equivalent to BS 459, Part 3, painted

Framed
143 40 mm softwood framed, glazed in two panes, painted
210 40 mm hardwood framed, glazed in two panes, polished

Frameless toughened glass
196 10 mm obscure armourcast hung with patch fittings in painted softwood frame, standard latch set

Flexible
403 In one pair, 8 mm black natural rubber panels fixed in painted steel framework with integral double action spring and standard vision panels

Metal
680 Polished anodized aluminium box section door and frame, glazed in two panes, overhead concealed spring
807 Dull polished stainless steel bonded to aluminium core door and frame, glazed in two panes, overhead concealed spring

Services

An analysis of a typical modern office block will reveal that this major element can, and frequently does, account for up to 34 per cent of total building costs (see Table 2 on page 41). Because of its increasing importance, somewhat due to the more widespread use of air conditioning, more quantity surveyors are being asked to measure the services element in their bills of quantities, instead of simply including large and somewhat inaccurate prime cost sums for them. To this end, many large private practices now have their own specialist quantity surveyors, although bills of quantities are not generally orientated to the measuring of large industrial engineering projects. In fact, unlike building works, there is no convenient, widely accepted standard method of measurement.

For the purpose of this section, and for the ease of this examination only, we have subdivided services thus:

1 Plumbing and waste
2 Heating and electrical
3 Air conditioning
4 Lifts and special services

Plumbing and waste

A hot and cold water system will be more economic, the more compact it is. For example, in multi-storey structures the kitchens and toilets on each floor should be positioned over each other to avoid long runs of pipework, which will add considerably to costs, and in the case of hot water pipes, only dissipates the heat. There are obvious exceptions where it is not possible to have a compact layout, for example in a school or hospital, and in projects like these the design team should be prepared for costs of services to be considerably in excess of 34 per cent of total costs. Therefore, the costs of the actual pipework, assuming a compact design has been chosen, would seem reasonably static. But the costs of sanitary ware can vary considerably, with specifications ranging from white glazed china with chromium plated taps, to Sicilian pearl marble and gold plated fittings. Here then, is an area where costs should be controlled, and where a possible solution to overspending may be found.

Heating and electrical

The heating installation of a building should be designed in sympathy with the structure into which it is to be incorporated and must be considered on two fronts, initial capital costs and running costs. This is because many of the available systems have widely different installation costs. The cost of fuel oil has recently risen sharply, whereas the cost of natural gas has increased at a slower rate. Therefore, with possible sharp rises in fuel costs, the correct choice of heating system will be seen to be very important.

When discussing plumbing costs it was said that it was the sanitary fittings that could make or break cost limits. In a similar way, when considering electrical installations the costs of electrical wiring and distribution are reasonably stable, but the cost of fittings, that is, lighting fittings, etc., can contribute up to 30 per cent of total installation costs. In the case of multi-storey buildings where each floor may be let to a different tenant, the architect should ensure that all lighting fittings are installed as part of the main contract and not leave it to each tenant to install his own.

Air conditioning

Air conditioning has, in recent years, become a more common client requirement, whether one is building for owner occupation or for letting. Even so it is mainly restricted to offices/hotels, etc. and many installations in the United Kingdom do not compare with the high standards demanded in North America. Until recently special air conditioning was required by all computer installations, but for the next generation of computers the purity, humidity and temperature of the air are not so critical and the general air conditioning plant may be used.

However, even allowing for this somewhat lower standard, if the decision is taken to include this facility in this country, the design team should be aware of the full financial implications, if the system is to function efficiently.

It is not simply a matter of allowing for the costs of plant and distribution; there are many other additional associated costs. Also the design philosophy of the entire building should be considered if air conditioning is to be provided, even down to such basic principles as to which elevations will receive direct sunlight and which shade.

The following factors can be said to influence the overall design and cost of air conditioning systems.

Distribution space   Assuming that there are five or six major types of air conditioning systems, it should be appreciated when designing that each type requires varying areas for ducting, trunking, etc., for example:

1   Fan coil system will require approximately 7 m² per typical floor of an office block
2   A dual duct system will require as much as 20 m² per typical floor

In addition it is common practice to conceal most of distribution ductwork within the suspended ceiling void, and again each system will require a different height between the suspended ceiling and the slab. In the two systems mentioned above, the height required varies from nothing to 450 mm, and this height will of course determine the overall floor to ceiling height, the overall height of the building, the enclosing ratio and overall costs.

Plant room space and heights   The choice of system will be the determining factor in calculating the size and height of plant room required, for example:

1   A fan coil system will require a plant room with an approximate total floor area of 325 m² with a clear height of 3.30 m
2   A dual duct system on the other hand will require up to 530 m² of plant rooms with 4.8 m ceiling heights

Window design, glazing and blinds   The windows will almost certainly have to be double-glazed, and either fitted with blinds, or glazed with sun reflecting glass to maintain temperature levels within the building.

Structural module and planning module

Building size

Depth of offices (rooms)

Noise insulation

Electrical leads and light fittings

Building location and orientation and resulting shade factors

Type of occupancy (that is, open plan or cellular offices)

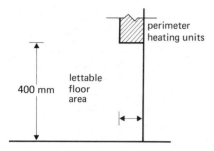

Figure 14

It should be pointed out that these factors vary for every building and cost studies should be undertaken to quantify the effect of them in any specific building project.

When using certain air conditioning systems the positioning of the installation can also be of financial importance, especially in buildings that are to be let to tenants. For example, perimeter heating units should be of sufficient height to permit say, a waste paper basket to be placed beneath them, as shown in Figure 14.

If it is not possible to do this then a prospective tenant may successfully argue that the area beneath the unit, albeit 150 mm wide is unusable and therefore does not warrant including in the net lettable floor area. Therefore, this detail, in a large office block with many floors, could substantially affect the gross/net ratio and the revenue.

Lift installation
Perhaps the two most commonly used forms of lift installation are electrical traction drive and oildraulic drive. The main advantage for the use of oildraulic lifts is that the ride is smoother and also the levelling at floors is much more precise which is essential for invalid cars, hospitals, etc.

However, the cost of installing the oildraulic lifts together with the necessary bore holes, is approximately double the conventionally driven lift. Maintenance costs are approximately equal for both, but because the oildraulic lifts have no large driving motor, the running costs during their working life are negligible.

As with an air conditioning installation there are many additional associated costs that should be allowed for when preparing cost limits for lifts. These costs range from additional electrical work to work done to motor rooms and general builder's work.

Finishes

This is indeed a vast element and also one which contributes a significant percentage to total costs (see Table 2 on page 41). Unfortunately, if a solution has to be found for overspending it is usually this element that the design team turn to first. Fortunately in recent years more effective cost planning has given the design team better alternative action. Finishes are generally incorporated in large quantities, and therefore, as with the external wall element, an increase or a decrease of a few pence per square metre affects overall costs considerably. In a book of this type there simply is not the space to discuss in detail the many thousands of different types of finishes. It may seem to the reader to be an easy way out, by saying that each project should be evaluated on its own merits, but nevertheless that is what should be done in this case.

For ceiling and wall finishes, dry lining has become very popular in the last decade as it is quick and simple to erect and allows work to progress rapidly. For floor finishes the recent price increase in oil based products means that 'cord type carpets', once thought a luxury, are cheaper to lay than vinyl floor tiles. Computer installations were mentioned briefly when discussing air conditioning, and it should be remembered that such an installation will require a more expensive special raised floor to accommodate the electrical installation.

In America fair faced blockwork has been extensively used in communal areas of flats and other public buildings and because of increasing brickwork and plastering costs, it is becoming increasingly common in this country too. However, to work successfully, the workmanship must be of the highest standard, otherwise the overall effect is catastrophic.

External works and drainage

As with the element substructure, this element is very difficult to evaluate. A multi-storey block on a restricted city centre site will have very little in the way of external works or drainage, whereas a large housing estate in a rural setting may need major sewer and landscaping works. Details of what is required are seldom available until during the contract period and even then the planning authority may alter the layout of the landscaping and the external works. Fortunately this element does not contribute a large proportion towards total costs. However, it does contain areas where a choice may have to be made between several alternatives after comparing initial and maintenance costs, for example fencing and road surfacing. Therefore for items like these, it may well be worth carrying out a costs-in-use exercise bearing in mind the disruption that may be caused if large areas of service roads have to be replaced regularly.

Industrialized building

If some distant planet were to observe our traditional building techniques they would think us a very primitive race, constructing the majority of our buildings by cementing one block to another. Industrialized building techniques have been tried and tested for many years, surprisingly without making any significant impact on the construction industry. Perhaps this is even more surprising when, as we have seen throughout this book, traditional methods are labour intensive and labour is a commodity that is rising steeply in price.

In traditional building the sequence of the work and the tasks performed are well understood and many organizational problems are mitigated by a relatively slow rate of building. Prefabrication in the form of staircases, door sets and trussed rafters, has been rather successfully assimilated into the process, but some attempts to introduce new techniques or greater pre-fabrication and a faster rate of building have failed, due to a lack of understanding of the factors which affect productivity on the site. If new techniques are to succeed, there is a need to provide a feedback of information from the site to the effects of innovation, so that design can be continuously modified to improve output.

At the beginning of this chapter, when discussing substructures, the *Finchampstead Project* was mentioned and during this project methods for collecting data of man hour requirements on site were developed and used by the Building Research Station. The technique which has been found to be particularly suitable for detailed studies is activity sampling.

With this technique one observer can study about 100 operatives. Briefly, in an activity sampling study the observer 'snaps' each operative on site at set moments of time, recording the work he is doing at the moment according to a defined code. The results shown in Table 3 have been obtained from the activity sampling at the *Finchampstead Project*. The total labour require-ments on site expressed in man hours per dwelling were 930, well below the average of about 1200 man hours for traditional construction.

Average man hour requirements per dwelling	
Distribution mains and service connections	60
Site development and external works	310
Substructure	70
Superstructure	90
Services	145
Finishes	255
Total	930

Table 3

Therefore, if the man hour requirements can be so drastically cut and yet industrialized systems still cannot compete costwise with more traditional methods, then the higher costs must lie with the manufacture of prefabricated components. The capital investment necessary for an industrialized system of building is almost twice that required for traditional methods. This is because many of the prefabricated components have to be made on purpose built factory type production lines and these initial costs are so great that, in the case of housing, it is generally accepted that the minimum number of houses is 50 before the 'break even' point is reached. Before leaving the costs that are associated with labour it is worth noting that although the total number of man hours is lower, a higher proportion of these hours is taken up by skilled erectors and fitters rather than traditional operatives.

Therefore, it would seem that industrialized building techniques can be economically used only under certain circumstances, viz.:

1   Low rise housing in schemes of fifty or more units are especially suitable, provided that there is sufficient time in the pre-contract period to set up the necessary organization
2   As so many of the 930 man hours per dwelling in the *Finchampstead Project* were taken up by site development and external works, about one third (310) hours, then the site should be as level and uncomplicated as possible
3   The site management staff should be familiar with the system and realize that any problems over constructional details, etc., have to be solved immediately and will not have the opportunity to 'come out in the wash'

It has been said that the use of industrialized building techniques and prefabricated components are bound to increase in the future because of the serious manpower shortages threatening the industry, but, as yet, apart from a few schemes, there seems to be little sign of increased use in these forms of construction in the next decade. However, complacency does not build houses and the Department of the Environment is aware of the present unacceptable time-lag when using traditional design processes, from the initial decision to build to people actually moving into new council homes. Prefabrication will reduce the construction time, as we have seen in the *Finchampstead Project*, but appears to be unacceptable because the associated costs are too high. A large housing scheme will, naturally enough, require a large site, usually bought with borrowed capital at high interest rates and the new houses will not start to produce an income until completion, which could well be years from the date the site was originally purchased. Therefore, with the aim of trying to reduce the total construction period the following procedures are a selection of the ideas that are currently being tried and tested.

Firstly, the Department of the Environment has developed a number of ideas based on the 'develop and construct' procedures. These procedures enable contractors with design skills and constructional techniques to participate in the design process in order to make schemes easier to build and to reduce the design, tendering and construction period while competition between contractors is still maintained. In the Property Service Agency's 'develop and construct' procedure the architect prepares a detailed site layout and selects dwelling types from a range of PSA standard plans based on metric house shells. The selected contractors tender on the basis of the layout and plans, a performance brief and specification clauses, their own method of construction and foundation drawings. The successful contractor then 'develops' that is, details the design into final working drawings to suit his own method of construction and finally carries out the construction. All drawings produced by the contractor have to receive the architect's approval before use, but the contractor bears the responsibility for design failure.

In the National Building Agency's 'design and build' procedure, the architect prepares a site layout using metric shells and selects contractors with ranges of dwelling plans to fit the metric shells. The contractors tender on the basis of the layout and their own plans, a performance brief and specification clauses.

4 Approximate estimating techniques

It is essential throughout the whole design process that the quantity surveyor has the techniques available in order that he can evaluate the scheme under consideration as accurately as the information he has available will allow. The main function of approximate estimating is for the quantity surveyor to provide a preview of the probable tender figure. He needs to do this for two reasons:

1 To make the client aware of his probable financial commitment as early as possible to avoid the waste of expensive resources
2 To let the architect and/or client know if the design is at all feasible

It is important to note that the first figure that the client hears is usually the one that he will remember, hence, it is equally important for the quantity surveyor not to overestimate as it is for him to underestimate. As already stated the estimate will only be as accurate as the information available and the skill of the estimator, however, as the design develops and more information becomes available the certainty of accuracy should increase. The accuracy of the estimate will be reflected in the amount of price and design risk that is incorporated into the estimate: indeed many surveyors call the price and design risk element the 'surveyors contingency sum'.

Basically, approximate estimating can be divided into two parts:

1 *Preliminary estimates* – these being to establish the broad financial feasibility of the project, for example the unit method

2 *Later stage estimates* – these being to produce a figure comparable with the lowest tender figure

The methods for use by the quantity surveyor are as follows:

1 Preliminary estimates
   (a) unit method
   (b) cube method
   (c) superficial method
   (d) storey enclosure method

2 Later stage estimates
   (a) approximate quantities
   (b) elemental estimating

By careful choice of which method to employ at a particular stage during the design of the project, the quantity surveyor should then be able to meet the aims and objectives of cost planning and cost control systems.

The unit method

The unit method is a single price rate method based upon the cost per functional unit of the building, a functional unit being, for example, a bed for a hospital building or a seat for a theatre. This method of estimating should wherever possible be restricted to inception and feasibility stages of the design.

The method is often regarded as a way of making a comparison between buildings in order to satisfy the team that the cost is reasonable in relation to other buildings of a similar nature.

The quantity surveyor should not use this method to estimate the cost of a specific building as no adjustments can be made to the single price rate. Very often the quantity surveyor will be asked to express his estimate in terms of a cost per unit, but this would normally only be at the request of the client, say a regional health authority.

The cube method

The cube method is a single price rate method based upon the cost per cubic metre of the building. This method of estimating should be restricted up to outline proposal stage of the design. Historically the method was used extensively between the wars but has now been almost totally superseded by the superficial method. At one time contractors kept what was called a cube book, in which they would record the cost of a project that they had undertaken and express that cost as a cost per cubic metre but this procedure has almost died out completely.

This method will give the surveyor some indication of the total cost of the project, but now its use is restricted to calculating heating requirements for buildings or for estimating fire insurance premiums. Where different parts of the building vary in function then the different functional parts should be quantified separately.

Its one advantage is that it is relatively quick and simple to calculate a total cost for the project, though its simplicity leads in some cases to inaccuracy as it is extremely difficult to assess the unit rate due to the large number of variables. The large unit quantity means also that a few pence error in the rate when extended can make a vast error in the final sum. The method itself fails to take account of any variation in plan shape, storey height or total number of storeys – each of which has an important influence on the final cost.

For example: a single storey car park with a volume of 600 m² will, using this method, equate to the same final cost as a three storey car park with a total volume of 600 m². As everyone knows this cannot be so.

Care should be taken when using historic data from previous projects to obtain a price rate, to ensure that one is aware of exactly what the rate includes – it is often necessary to make some addition for lift installations, etc.

The calculation of the volume is, however, subject to rules of measurement in order to ensure, like the Standard Method of Measurement, that all estimates are calculated on the same basis. These rules are outlined as follows:

1   All measurements are from external faces of external walls.
2   The height for a pitched roof building is taken from the top of the foundation to a point midway between the ceiling and the apex of the roof, where the roof space is unoccupied. If the space is occupied, as is the case with dormer rooms, then the height is taken again from the top of the foundations but to a point three-quarters from the ceiling to the apex of the roof.
3   The height for a flat roof building is again taken from the top of the foundation but then to a point 600 mm above the roof structure.

Example
Refer to Figure 15.

Volume of office accommodation:

$$25 \times 7 \text{ m} \times 7 \text{ no.} = 1225 \text{ m}^2$$
$$10 \times 3 \text{ m} \times 6 \text{ no.} = \underline{\ \ 180 \text{ m}^2}$$
$$\qquad\qquad\qquad\quad 1405 \text{ m}^2 \times 3 \text{ m} \quad = 4215 \text{ m}^3$$
$$25 \times 7 \text{ m} \qquad = \overline{\ \ 175 \text{ m}^2}$$
$$10 \times 3 \text{ m} \qquad = \underline{\ \ 30 \text{ m}^2}$$
$$\qquad\qquad\qquad 205 \text{ m}^2 \times 0.6 \text{ m} = \underline{\ \ 123 \text{ m}^3}$$
$$\qquad\qquad\qquad\qquad\qquad\qquad\qquad 4338 \text{ m}^3$$

Cost of block $4338 \text{ m}^3 \times £101 \text{ per m}^3 = \underline{£438,138}$

The superficial method

The superficial method is a single price rate method based on the cost per square metre of the building. This method of estimating is again restricted up to outline proposal stage of the design. In practice, this method is probably the most popular of the preliminary estimating methods. Its major advantage is that most published cost data is expressed in this form, whether it be in historic or present day form. The method is quick and simple to employ, though once again it is imperative to use data from similarly designed projects.

The superficial method has the advantage that it is meaningful in its

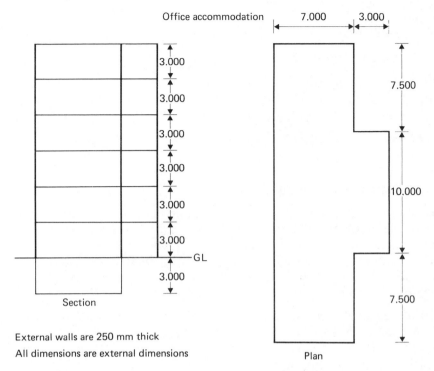

Figure 15

concept of measurement (the square metre) to both the client and the architect. The client can most likely visualize a building with 1000 m², but will have great difficulty with the concept of cubic metres. The same size building here then would have approximately 3000 m³ – which to most people seems a lot larger.

However, this method also fails to take account of any variation in plan shape or storey height. Therefore, it is essential to obtain data from a project with a similar shape and height. Care should be taken when choosing the appropriate rate to ensure that it only includes for the specification that is on the proposed project – very often, the rate will not include for such things as lift installations or abnormal foundations.

The rules for calculation are much more simple, and therefore, the method is quick and easy to employ:

1 All measurements are taken from the internal face of external walls. No deductions are made for internal walls, lift shafts or stairwells – the concept here being that the cost of forming those openings in a floor is counteracted by including for the cost of the floor to those openings.
2 Where different parts of the building vary in function, then the areas are calculated and quantified separately.

Example
Refer to Figure 15.

Gross floor area of the office block:

24.5 × 6.5 m × 7 no. = 1114.75 m²
 9.5 × 3 m   × 6 no. = 171.00 m²
                      1285.75 m²

Cost of block   1285.75 m² × £375 per m² = £482,156

The storey enclosure method

The storey enclosure method is a single price rate method that attempted to overcome the problems of all the other preliminary estimating methods – it attempts to take account of variations in plan shape and storey height.

The method was introduced following a study by the RICS in 1954 under the chairmanship of W. James. Unfortunately the method was never totally adopted by the profession for a number of reasons:

1   Far more calculations are required than for the other methods
2   Historic data in the form of rates are not readily available
3   It still retains the disadvantage that the concept of 'storey enclosure units' is not meaningful to the client or the architect

The basic theory of the storey enclosure unit method is that it consists of measuring the area of external walls, floors and roof areas (effectively 'enclosing' the building) and to those areas, multiplying by an appropriate weighting factor thus resulting in a number of storey enclosure units.

As would be expected, this method involves the quantity surveyor in a lot of very tedious calculations which result in a meaningless unit of measurement.

The rules for calculation are as follows:

Floor areas
These are measured from the internal face of external walls, and are subject to the following weightings:
Basements   × 3
Ground floor × 2
First floor   × 2.15
Second floor × 2.30
                  and *add* 0.15 for each successive floor

Roof areas
These are measured to the extremities of the eaves and are measured as the area on *plan* whether they be pitched or flat. There is no weighting applied to the roof area, that is, × 1.

External wall areas
These are measured on the external face of the walls and are subject to the following weightings:
Basement wall area (basement floor to ground floor level) × 2
Above ground wall area (ground floor to ceiling of top floor with no deduction for openings) × 1

The number of storey enclosure units is then multiplied by a single price rate – additions should then be made for:

1　Services
2　Abnormal foundations
3　External works and external services

Example
Refer to Figure 15.

Floor areas			Storey enclosure units
Basement	24.5 × 6.5 m × 3	=	477.75
Ground	(24.5 × 6.5 m + 9.5 × 3 m) × 2 =		375.50
First to fifth	187.75 m² × (2.15 + 2.3 + 2.45 =		
	+ 2.6 + 2.75)		2299.94
Roof area	(25 × 7 m) + (10 × 3 m) × 1	=	205.00
Wall areas			
Basement	64 m × 3 m × 2	=	384.00
Ground to roof	70 m × 3 m × 6 no. × 1	=	1260.00
	Storey enclosure units		5002.19
Cost of block	5002.19 units × £103.28 per unit =		£516,626

Note: The unit cost includes for lift installation and external works.

Approximate quantities

This is regarded as the most reliable method that the quantity surveyor can employ, provided that there is sufficient information for him to work from. Because of this limiting factor, this method of estimating is categorized under the heading of *later stage estimating*.

It is a popular method with the quantity surveyor because it adopts, as its name suggests, the traditional methods of taking-off for the preparation of the basic data and then involves applying the use of composite rates from bills of quantities for its pricing.

It is a reliable method of estimating as it is easy to make adjustments for the variance in the quantity or quality of materials involved. This method also makes an automatic allowance for plan shape height and size. If the estimate is prepared too early, before there is sufficient information available, then the quantity surveyor could produce an inaccurate estimate. Very often, however, by the time the quantity surveyor has sufficient data to begin preparing the approximate quantities the estimate is no longer required for the client – in other words, it is often left until the drawings are being prepared for the bill of quantities proper, by which time it is too late. The very nature of the estimating method employed renders it time-consuming for the quantity surveyor.

The rules of measurement are simple for most quantity surveyors as they involve grouping together items relating to a sequence of operations and relating them to a common unit of measurement. Composite rates are then built up from data available in the office and applied to the units of measurement in order to arrive at a cost for that sequence of operations. All measurements are taken as gross over all but the very large openings. Windows and doors are priced as a complete unit separately.

Example
Excavate foundation trench 1 m deep; level and
compact the bottom of trench; earthwork support;
backfill; disposal of surplus spoil; concrete
foundations 300 mm thick; cavity brickwork to
150 mm above ground; bitumen based d.p.c.;
facing bricks externally.   50 linear metres @ £52.56 per metre

Elemental estimating

This method was evolved from the storey enclosure method, which considered certain major elements separately. The method started its life in the form of the cost analysis, which evolved to provide a method of isolating the cost differences between buildings, but has now provided the profession with a valuable source of historic cost data. With this data, the quantity surveyor is able to produce an estimate in elemental form thus overcoming all the major disadvantages of the preliminary estimating methods.

The basis of the estimate is to consider the building, as its name would suggest, in elements. The form of construction of the element, its material and labour content, the quantity involved are all compared with an element of an analysed building. The elemental rate is then adjusted to take account of any necessary quantity or quality alterations, and multiplied by its relevant unit quantity to arrive at the cost of the proposed element.

The unit of measurement can either be the gross floor area, in which case the cost per square metre of gross floor area is taken, or the element unit quantity is multiplied by the element unit rate to arrive at an elemental cost. The choice is generally left to the quantity surveyor, though it is easier if there are a lot of adjustments to be made, to evaluate the cost using the element unit quantity and element unit rate.

This method has the advantage that it can be used at all stages of the design again provided that there is sufficient information available. The client should be capable of understanding the estimate. It is a very reliable method of estimating which means that comparisons can very easily be made thus enabling the architect to obtain any likely cost implications during the early stages of the design. This being so, however, the quantity surveyor will need a great deal of skill to be able to adjust the elements correctly to take account of any changes in specification which could prove to be very time-consuming.

As already stated, this estimate is made in the form of the major elements of the building. A list of the elements used in the majority of cases is given in Appendix B.

5 Construction cost indices

The cost of any building design is determined, primarily, by the cost of the labour and materials involved in its erection. Any variation in the cost of either of these basic factors will influence the cost of an item of work.

Cost indices, therefore, are an attempt to measure those price variations that occur between tenders obtained at different times. The quantity surveyor is then able to study the indices together with any predicted future cost trends, albeit a rise or a fall, and any regional variations, in order to facilitate 'price adjustments' to historical cost information.

Generally, the client will need to know any change in price which affects him on a functional basis, that is, how much more expensive is this factory unit going to be compared with the one built two years ago? This would mean that the index would have to combine changes in quantity with changes in price. However, the only index providing that sort of information is the *average selling price of new houses* published by the Building Societies, and that index will also include for the cost of the land. In all other cases then it will be a case of treating the changes in price separately from any changes in standard or quantity.

The most common forms of indices are:

1   Price indices (measuring the price to the client)
2   Cost indices (measuring the cost to the contractor)
3   Location indices
4   Maintenance indices

Price indices

As already stated, these indices will measure the changes in price paid for by the client. This form of index is the one most commonly used by the quantity surveyor, as he is acting for the client and will be reporting any changes in cost to the client.

It is primarily used for updating historic cost data for use on present day estimates. If the index is based on a fully fluctuating contract, then obviously the index will not take account of the *total* price, but it will exclude any payments for fluctuations. The index may be based on the cost of tenders – in which case it will become a tender based index – or on any other data that the quantity surveyor has available to him relating to the total price.

Price indices can be obtained from any of the following sources:

1   The *DQSS Tender Price Index*   This is published quarterly in the *Chartered Quantity Surveyor* and originated as an index for general building work calculated by the former MPBW. The index measures the changes in unit rates quoted by tenderers, but does not make any allowance for the variation of price additions.

2   The *BCIS Tender Price Index*   This is identical to the DQSS index but also deals with both public and private sector work and includes housing. The tenders are included in the index for the quarter in which they are submitted.

3   The *Davis Belfield and Everest Tender Price Index*   This index is published in the *Architects' Journal* and is based on a similar principle to the BCIS index. The index also includes private and public sector work as well as housing.

4   The *Public Sector Tender Price Index*   This is published by the DoE in the form of *Housing and Construction Statistics*. It is a combination of the DQSS Tender Price Index with the similar indices for other government departments and local authority work. It contains an assortment of general indices including costs, prices, output and performance.

5   The *Price Index of Local Authority Housing*   This is also published by the DoE *Housing and Construction Statistics* but only covers local authority housing work. It is a Laspeyres index based on a standard weighting of about twenty-three items which usually occur in a housing bill of quantities.

6   The *Roads Construction Price Index*   Also published in the *Housing and Construction Statistics* covering new roads construction which are controlled by the Department of Transport.

7   The *Out-turn Price Index*   This is also published by the DoE *Housing and Construction Statistics* and has officially replaced the *Cost of New Construction Index* which is no longer recommended. There are separate indices for public sector and private sector work that are divided into categories such as housing, industrial building, etc. Each index measures the price the client is currently paying for that type of building. It consists of the appropriate tender price index plus an addition for the average amount currently being paid by the operation of the formula variation of price conditions.

Cost indices

These generally aim to measure changes in cost to the contractor, not the price to the client. Therefore, they ignore any changes in profit levels and overheads.

Since they are usually based on published wage rates with a standard allowance for plus rates, etc., they tend to ignore any changes in productivity, discounts for purchase of materials, etc., therefore, they effectively measure changes in the 'notional' rather than the actual costs.

The primary purpose of these indices is to determine the amount to be paid to contractors under the operation of variation of price clauses in the contract. However, it is obviously a simple matter to analyse a sample of any particular building type or types into its 'works categories' and apply the variation of price indices each month to form a Laspeyres-type index.

The basic source of most present day cost indices is the set produced by the PSA for the operation of the variation of price. There is a set for building, civil engineering and for specialist subcontractors.

The make-up of each index is published in the PSA's *Guide to Indices*. The labour and materials weightings and the constituents of the labour indices are determined by a working party of the National Consultative Council representing both sides of the industry.

Location indices

These are generally compiled from the tender price index and attempt to make some allowance for the variation due to location.

The BCIS annually publish a set of location indices which cover most parts of the country. These are then considered when attempting to make an assessment of the variation in price of a proposed project where the location is vastly different from that of the analysed project.

Maintenance indices

A price index for the value of maintenance work only, is prepared jointly by the DQSS and PSA and is published quarterly in the *Chartered Quantity Surveyor*. It measures the average price being paid for maintenance work under the Measured Term Contracts for PSA buildings, and is compiled from the Addenda A percentages (these being the percentage additions made to the PSA Schedule of Rates for Measured Term Contracts compiled by applying the formula variation of price indices to the same sample of orders) and the tender percentage additions as applied to a large sample of orders for maintenance work.

It is important to note that even with the assistance of the many varied forms of indices, the forecasting of future building costs can be a somewhat hazardous occupation. If you take a look at any set of indices published it very soon becomes obvious that the movement of prices can be very erratic.

For some years now there has been concern regarding the degree of reliability of the available building price indices for the following reasons:

1   There is a notable lack of consistency between the various published indices

2   There is a general failure to indicate the possible future movement in building prices
3   The wide coverage of the majority of indices shows little regard for a specific locality or building type

Despite this, the system of indexing is the best system available to the quantity surveyor at the present time and remains relatively consistent in its predictions.

Application of the indices

Building cost indices are often used to bring a tender figure for a previous similar contract up to current prices in order to use it for comparison purposes as part of the cost planning process.

It is important then to understand clearly what items are included in an index and exactly what date the contract was indexed from (this becoming the *base index*). Having established the *base date* it is then possible to update from that base to the current (or even future) date using the index system.

Most government statistics are re-based on current weightings at five or ten year intervals. Therefore, when comparing the movements of two indices, it is important to see that any linkage is made at the correct time.

It may also be necessary to make some allowance for increased costs during the period from the preparation of the estimate until the probable tender date. This sum is often referred to as the 'price and design risk' and calculated as a percentage based on the likely time period and the current price trends, taking into account the market conditions.

Example
Analysed tender for a steel-framed office building (firm price contract)
Tender figure   = £160,000
BCIS index at tender date (base)   =   153
Present BCIS index   =   190

Estimate for project of similar design

$$\text{Value of project} = £160,000 \times \frac{190}{153}$$
$$= \underline{£198,500}$$

6 Cost planning during the inception and feasibility stages

Inception stage

At the inception stage of the design sequence, the task of designing a building that will conform to the client's requirements is a matter between the architect and the client; because of this, any discussions as to the form and the construction of the proposed building can only be at a very general level. It will not be until the other members of the design team have been appointed that the detailed discussions can begin. The idea, if indeed it still exists, of the building client being a single wealthy individual, with only a vague idea of what his requirements are, should be dispelled. The majority of modern clients will be the representatives of large organizations or companies, with a rigidly fixed budget and overall annual expenditure. The financial targets for future building programmes are generally fixed in advance and cannot be exceeded. It is for this type of client body that cost planning and cost control can offer a necessary service. Assurances will be required early on in the design development that an idea which appears to be perfectly feasible on paper will turn out to be feasible in practice and not run into any financial difficulties.

One of the reasons for the increasing importance of cost planning in recent years is the fact that construction methods are becoming increasingly more complex. The combination of new materials that it is possible to use, in any particular design solution to a constructional problem, seem endless at first glance to an untrained eye. Indeed, so rapidly are new materials and processes being discharged on to the market that comprehensive directories have had to be compiled in order that design team members can keep up to date with the new techniques. Examples of such directories are the *National Building Agency*, the *Barbour Index*, and *'Building' Commodity File*, which are all information services intended to keep design team members fully informed of the types of materials that are available for their use.

No longer is it possible for the quantity surveyor to 'spirit' figures from out of the air and quickly estimate the cost of a particular design. Instead, the use of cost planning and cost control can and should produce an accurate first estimate of the cost, and then also prevent the design and the cost of the project from running out of hand, when unfamiliar techniques and untried materials are employed, by directly linking the cost of the building to the design, with the use of an elemental cost plan.

After the initial meeting with the client, the architect will be able to appoint the professional assistance that he considers necessary for the successful completion of the project. There has been a reference earlier in this book to the appointment, at this stage, of specialist subcontractors. The principal reason for this is that a great deal of difficulty is experienced by most quantity surveyors when calculating cost targets for the services elements, which are usually installed by the specialist subcontractors.

Taking as an example elements numbers 5E and 5F, the elements associated with heating, it is obviously unrealistic to apply to these elements a common measure, such as the square metre, when calculating their relative costs, due to the vast amount of systems that may eventually be installed, all having different capacities and outputs. The design of the building and the materials used in its construction will have an effect on the total heat loss, the size of heater and the output of the system that will be required to give the correct heat levels.

In the case of element 3B, floor finishes, the cost of this element is expressed in two ways – as a cost per square metre of the gross floor area and as a cost per element unit rate. The gross floor area is as expressed by the Building Cost Information Service as:

1   The total of all enclosed spaces fulfilling the functional requirements of the building, measured to the internal face of the enclosing walls
2   Includes the area occupied by internal partitions, columns, chimney breasts, internal structural or party walls, stairwells, lift wells, and the like
3   Includes lift, plant, tank rooms and the like above the main roof slab
4   Sloping surfaces such as staircases, galleries, tiered terraces and the like should be measured on the flat

As this element does not have as many variables as the heating element, it is possible, in the majority of cases, to use the information that is contained in the cost analysis in the preparation of the cost plan using only the standard adjustments for price level, quantity and quality. However because the heating element can be infinitely variable in its type and size, it is not accurate enough to adopt the same technique in the calculation of the cost target.

It is for this reason that it is good policy to invite as many specialist subcontractors as possible who are involved in the contract, to submit their estimate early in the design sequence. As over the past few years the whole field of environmental science has become the 'lot' of the specialist, it is only reasonable to leave the preparation of this part of the cost plan to him.

As stated earlier, elements such as the floor finishes element can be expressed in two ways – a cost per square metre of gross floor area and a cost per element unit quantity. Of these two pieces of information, the element unit rate is probably the more useful, and yet, ironically, it is the one piece of information that is usually missed from certain published cost analyses. The element unit quantities for each individual element are not calculated haphazardly, but in strict accordance with a rigidly defined set of rules, which unfortunately differ with the publisher of the cost analyses. The inclusion of the element unit rate in the cost analysis greatly simplifies matters, as adjustments for changes in cost, due to differences in gross floor areas, are automatically made.

Example (Figure 16)
It is proposed to prepare a cost plan for Building A based on the information that is contained in the cost analysis for Building B.

Building A: gross floor area 100 m²;   roof area 100 m²
Building B: gross floor area 200 m²;   roof area 100 m²

An identical form of roof covering and construction is to be used for each building. Assume the cost of the roof covering to be taken as £5.00 per square metre, this would give a cost per square metre of gross floor area of the analysed Building B of:

$$\frac{100 \times £5.00}{200} = \underline{£2.50}$$

If in the preparation of the cost plan for Building A, the rate of £2.50 is used, unadjusted for differences in floor area, although the cost of the roof in

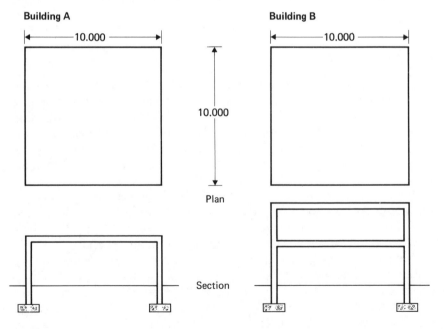

Figure 16

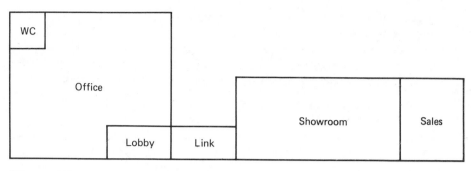

Figure 17

both buildings is the same that is, £500.00, it would produce an answer of:

$$100 \times £2.50 = \underline{£250.00}$$

Therefore, although it is possible to make compensating adjustments to the rate of £2.50 and obtain the correct answer, the piece of information that is of real use to the quantity surveyor is the element unit rate, which in this case is £5.00 per square metre.

Feasibility stage

With the design team members now appointed and a good communication system established between them, the task of preparing the feasibility report can begin. Above all others, this is perhaps the most important stage in the early design sequence; accuracy is essential, because a careless error at this stage could involve the whole design team in much embarrassment later on when the full cost plan is prepared.

At this stage the quantity surveyor has to prepare the first estimate by the interpolation of all available relevant cost data. After the initial brief with the client, the architect will be able to establish the basic criteria he will have to follow when producing his sketch plans and design. The degree to which the architect will be able to crystallize his ideas at this stage will, to a large extent, depend on his experience and professional ability, but the broad principles of the project must be agreed between the architect and the quantity surveyor before a serious attempt at an approximate estimate can be made.

The quantity surveyor should try to notify all those concerned with the project that the following information should be made available before the feasibility report estimate can be prepared:

1   Sketch plans and elevations. The drawings need not, and indeed cannot, be very detailed, but should try to indicate wherever possible the type of accommodation to be provided in particular areas. See Figure 17.

2   A general indication of the overall standards that will be required. This may be by reference to another project.
3   The location of the proposed site in order that any adjustments for abnormal site conditions may be made. This factor may affect the allowance necessary for the preliminary and contingency items.

It is now a wise precaution to establish with the client the amount, if any, he has used in his user requirements for circulation space, for example cloakrooms, corridors, etc. The client may have expressed his requirements only in terms of functional area and any necessary allowances will have to be added to this. Another point to remember is that all the different circulation areas may have different types of construction and finishes, and therefore, differing costs per square metre.

Interpolation

This is the process of inserting into, or deducting from, the cost analysis figures various allowances in order to arrive at the figures that can be reliably used in the production of a preliminary cost plan for a similar project.

Armed with the information listed above, the quantity surveyor can now prepare the *feasibility report estimate*. The first stage in this procedure is to examine all the relevant cost analyses in the library of cost information. This may take the form of; the BCIS data, bound together in their own folders, the office's own priced bills of quantities, preferably prepared in elemental form, or the information that has been extracted from the various weekly and monthly journals and filed together into groups of similar projects or buildings.

From the range of cost information that is available, one cost analysis should be chosen, the classification (for example offices, single storey factories, etc.) and construction of which are similar to that of the proposed building. It is left to the quantity surveyor to choose the cost analysis and basically the choice should be made on the basis of similarity between the major elements, for example those which contribute the largest percentage to the overall cost of the building:

1   Substructure
2   Frame
3   External walls
4   Upper floors

Having chosen the cost analysis, it will then most probably be found that although there is a similarity between the major elements of the two buildings, there will still be a number of small differences for which allowances will have to be made.

It would indeed be unprecedented good fortune to discover an exactly similar building on which to base the estimate. So, having inspected the two

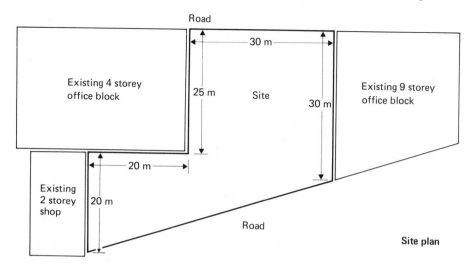

Site plan

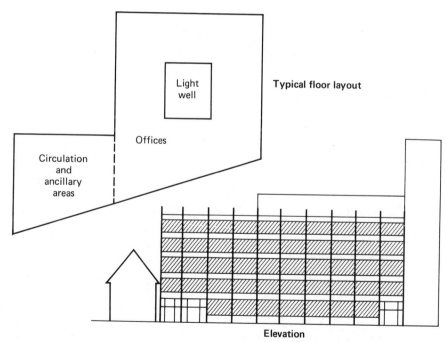

Typical floor layout

Elevation

Figure 18

projects to uncover the differences that exist between them, the quantity surveyor can make allowances for them.

Practical example 1: cost limit calculation

Throughout the remainder of this chapter, reference will be made to the cost analysis contained in Appendix C, which has been prepared and presented in accordance with the BCIS Standard Form of Cost Analysis. In the examples that follow in this and the next two chapters, it will be assumed that a cost plan is to be prepared for the same client, for a similar project, but in a different part of the country. Details of the proposed project are shown in Figure 18.

There seems to be little point in listing all the possible differences that may be found between cost analysis information and details that may be required for a proposed project. For there are so many that the remainder of this book would be taken up in giving examples of each. However, this example will endeavour to take students through the procedure the quantity surveyor would adopt in order to complete the final cost plan. This will be done through to its completion by way of the practical examples 1, 2 and 3 encountered here and in the next two chapters. At this stage the amount of information that is available is still very limited, and therefore it would be a somewhat pointless operation to prepare a feasibility report estimate on an elemental basis.

It has been decided to prepare the first estimate for the Office Development Company on the basis of the following:

1   Office accommodation
2   External works
3   Preliminaries and contingencies

The cost of each of these will be dealt with separately when preparing the estimate for the proposed building.

The architect has formulated one basic design solution that complies with the client's requirements. The objectives of the first estimate are two-fold: firstly to provide the client with a recommendation as to which of the designs to adopt, and secondly to give him the cost limit amount and feasibility study for that design. It has been decided, taking into consideration the information available to the quantity surveyor, that the best method of approximate estimating to use for this project is the superficial method. The reason for this is that the quantity surveyor has found among the historic cost information in the office, a cost analysis for an office block of a similar standard of finish, construction, and size. (This analysis is the one detailed in Appendix C.)

The order in which the adjustments are made differs from stage to stage, and at feasibility stage the first adjustment that is usually made is for differences in quantity. Quantity is not used here in the strict sense of the word, as it is used later on when it refers to the actual amounts of element unit quantity, or floor areas. Instead, it has been used to describe items which can broadly be said to fall into the category of adjustments required for differences in quality.

In the case under consideration, these items are:

(1)   The cost analysis contains items that are not intended to be included in the plan for the new building, and the cost of these redundant items will have to be deducted before proceeding: for example, in the cost plan project it has been discovered that the client wishes to install the heating system himself, as he has fortunately recently purchased a number of wall-mounted electric convector heaters. Therefore, the cost of these elements, numbers 5E and 5F, do not have to be included in the overall cost of the building. Owing to this, it is necessary to deduct from the cost analysis figure, the allowance that was included for heating. For this operation the cost analysis figure has to be subdivided into three groups of costs.

Office	£882,336
External works	£ 66,700
Preliminaries and contingencies	£111,280
	£1,060,316

Now the adjustment to the office cost has to be made for the heating allowance.

Cost of office	£882,336
Less allowance for heating	£118,436
	£763,900

A further deduction is necessary to the external works allowance, for the cost analysis contains a sum of £19,500 for an electricity sub-station and associated works, which will not be required in the proposed project.

Cost of external works	£66,700
Less allowance for sub-station	£19,500
	£47,200

It has been decided to incorporate the preliminaries and contingencies untouched; therefore, at this stage there is no adjustment to be made.

(2)   The cost analysis does not contain items that are intended to be included in the cost plan for the proposed building. An allowance will have to be made for this; for example, in the cost analysis there was no provision made for the cost of display cases for the company's products. The client has expressed a wish for the cost plan figure to include an allowance this time for wall-mounted, glass-fronted display cabinets. The price of these units will be obtained from the manufacturer and will be current at the time of the preparation of the cost plan. It would, therefore, be incorrect to add the display cabinet allowance of £20,972, as only costs from the same price level can be added or deducted directly to or from each other, and the later figure

will have to be updated for differences in price level before proceeding further.

The second major adjustment that has to be made is for differences in price level. The information that will be obtained from the cost analysis will be outdated; that is, there will almost certainly have been substantial rises in the general level of building costs from the time of the preparation of the cost analysis and the preparation of the cost plan. The quantity surveyor will make his adjustments here, as explained in Chapter 5. The adjustments for differences in price level can now be made. The cost index at the time of the preparation of the cost analysis was 100, and the index at time of preparation of the cost plan is 202. Therefore, to update the cost analysis information it will have to be multiplied by $\frac{202}{100}$, therefore:

Office portion £763,900 $\times \frac{202}{100} = \underline{£1,543,078}$

Now that the cost has been adjusted, the allowance for the display cupboards can be added.

Updated cost from above	£1,543,078
Cupboards	£ 20,972
	£1,564,050

The external works allowance and the preliminaries and contingencies can be updated and added to the cost of the office portion.

External works £47,200 $\times \frac{202}{100} = \underline{£95,344}$

The total cost is therefore:

Office	£1,564,050
External	£ 95,344
Preliminaries and contingencies	£ 224,786
	£1,884,180

This total cost must now be divided by the gross floor area of the analysed building to obtain a cost per square metre.

$$\frac{£1,884,180}{5000} = \underline{£376.84 \text{ per m}^2}$$

At this stage it is not possible to use an omnibus element unit rate. The gross floor area of the proposed building, including circulation space, etc., has been calculated as 5838 m². Therefore, to determine the total cost of the building:

£376.84 $\times$ 5838 m² $= \underline{£2,199,991.90}$

Other important factors that affect the overall price level of a building contract are:

1　The location of the proposed project. It is necessary to know in which part of the country the proposed building is to be constructed, because it is obviously more expensive to build in London and the Home Counties than it is in, say, Lancashire, and an allowance will have to be made accordingly, usually on a percentage basis. While dealing with the question of location it may become apparent to the reader that it is also much more expensive to build in the centre of a large town or city rather than the suburbs, wherever the location may be, and allowances for this are included in the preliminary item.

2　The type of contract that is used. Generally one can say that a tender figure obtained from a negotiated contract will be more expensive than a contract that is based on competitive tenders.

3　The market conditions at the time when the tenders were sent out. For example, at a time when work for the building contractor is scarce, then he may well be prepared to cut his profit margin to a bare minimum and submit a tender figure lower than usual, to ensure continuity of work and full employment for all his men. On the other side of the scales, however, in times when work is plentiful, and the job book is full, the building contractor may only be tendering for the contract so as not to be seen as unreasonable to the quantity surveyor and the architect. Obviously under these circumstances the builder will not submit such a low priced tender as before.

In order that all these conditions may be taken fully into account, the quantity surveyor when he prepares his estimates, should refer to the cover sheet of his analysis (see Appendix C), which contains information on location, market conditions, contract particulars, etc.

At this stage, because the information is so scant concerning the specification and overall standards, it is not really realistic to make adjustments for differences in finishes and standards.

Because the cost plan is based on a cost analysis that has a certain set of fully described finishes it is possible to make it clear to the client that his new building will have a general standard of finishes, similar to those described in the cost analysis. This may be pointed out to the client, before the quantity surveyor starts to work on the preparation of the feasibility report, in order that the client has an opportunity to express his dislike at any of the general standards offered. Alternatively, a summary of standards of finishes can be given to the client at the presentation of the feasibility report. This then is the final adjustment that may be necessary and will be referred to throughout this book as adjustments for differences in standards and finishes.

To the target cost must now be added allowances for:

1　Price risk
2　Design risk

Price risk

There has already been one adjustment for the differences in price levels, that is, adjustments of the cost analysis figure by multiplying the appropriate figure by an index. Another adjustment is also necessary for allowances for the rise in the price levels between the preparation of the cost plan and the date that the job goes to tender. It can be quite common for this period to be from six to twelve months. During this time the prices could have risen by 15 per cent, an increase that would cancel out all the careful adjustments that had been made earlier and make the cost plan figure unacceptably inaccurate. The exact amount that should be allowed will depend upon the length of time involved and the rate by which labour and material prices are rising. In the example under consideration, the estimated time between the feasibility report estimate and the *tender action stage* is thought to warrant an allowance of 3 per cent. It is also now possible to make an allowance for differences in regional pricing. The cost analysis is for a building that was constructed in North Buckinghamshire, the cost planned building is to be erected in Telford New Town, and in this case it is thought that an allowance is not required, as the general level of building prices is approximately equal in both areas.

Design risk

An allowance must be made for uncertainties in the (as yet) far from finished design. The amount will depend upon the nature of the project; for example, in preparing the cost plan for the Mosque in Regent's Park, the amount allowed for the design risk would be greater than for a three storey office block in Shepherd's Bush. In the project under consideration, which is a reasonably straightforward building, an allowance of 1 per cent has been adopted. The target cost now reads:

Design target cost	£2,199,991.90
Add price and design risk 4%	£ 87,999.68
Total target cost	£2,287,991.58
say	£2,290,000.00

During the preparation of the feasibility report estimate, the quantity surveyor will have to be in constant communication with the other members of the design team. When the report is completed and all the information from the various members has been correlated by the architect, it will be presented to the client at a meeting of all the people concerned. This is good practice, as discussion can take place concerning any unsolved problems; for example the quantity surveyor may find that the estimated cost of the project is greater than the client's target cost.

If it is decided to proceed, then the next stages in the design sequence are *outline proposals* and *scheme design stage*, as outlined in Chapters 7 and 8.

7 Cost planning during the outline proposals and scheme design stages

Outline proposals

With the acceptance of the feasibility report by the client, the design team can proceed to the next stage in the design sequence, outline proposals.

According to the various professionals whose findings collectively formed the feasibility report, the project seems to be trouble free from the technical and practical viewpoints and as the design sequence progresses, the amount of information available should be constantly increasing. Even so, at this stage ideas have not been crystallized sufficiently to warrant a cost plan being prepared on an elemental basis. It may be possible, however, under certain circumstances to make a first allocation of cost over the groups of elements within the cost plan:

1   Substructure
2   Superstructure
3   Internal finishes
4   Fittings and furnishings
5   Services
6   External works

It should be emphasized that this operation can produce rather inaccurate results, due to the fact that even though the two buildings are equal, there is no guarantee that the costs will have been distributed in equal proportions over the groups listed above.

There are a variety of reasons why the cost allocation of two similar buildings should not be comparable. For example, the builder's estimator may have deliberately 'front loaded' the tender (that is, inflated the rates for the early work in the bills of quantities, for example site clearance, substructure, etc.), making compensating reductions in the later trades. The principal reason behind this is that the builder may accrue an early working capital at a time when his cash reserves are low. Unless spotted when the bills of quantities are analysed, the inflated information will be recorded and make the process of trying to reconcile the cost allocation on this project with a similar building, where the pricing has been carried out in a conventional manner, very puzzling. This argument may equally well be applied later on, when the full elemental cost plan is prepared using cost analysis information. During the scheme design stage, costs are compared at a more intimate level and therefore it is possible to spot obvious discrepancies more easily than is possible at this stage, where the price allocated to a group of elements can hide many anomalies. Also when dealing with such large groups of elements it will only be possible to deal with costs per square metre of gross floor area, which, as stated previously, are apt to produce misleading results.

In the case being considered it has been decided that a cost allocation at this stage will be unnecessary, and indeed with very little more information available than previously, it would seem an unnecessary and costly operation to produce more drawings and allocate expensive resources in preparing an estimate. This is not to say that under ideal circumstances some design teams, especially when dealing with much larger projects, may not find this group allocation extremely useful, but even so, it could perhaps be postponed until the scheme design stage, when more information is available.

Scheme design stage

Now the full elemental cost plan is to be prepared, because the amount of information that is available to the quantity surveyor should enable him to allocate the target cost much more accurately than it has been possible to do previously. The information that is required to prepare a cost plan is:

1   A cost analysis of a similar building. The quantity surveyor will already be in possession of this, as the cost analysis used at this stage will be the same one that was used in the preparation of the feasibility report estimate.
2   Sketch plans and elevations, which will enable calculation, usually by scaling direct from the drawings, of the various element unit quantities and ratios for the cost plan building. At this stage it is important that the drawings should contain as much accurate, not assumed, information as possible, for the maxim applies, 'the less accurate the information, the less accurate the cost plan figure'.
3   It should now be possible to produce a more accurate and detailed schedule of proposed finishes. Nothing too detailed, but sufficient for the quantity surveyor to be able to justify making an adjustment for different standards and finishes in the two buildings. An advantage to be derived from starting the overall planning of a project at the earliest possible date is that detailed information will be available to the design team much earlier than if the problem had not been given thought until this stage, or even later. If the type and overall standard of finishes had not really been considered until this stage, due consideration may not have been given to them and hasty decisions may be made.

At this stage the cost of the proposed project is considered on an elemental basis, and any necessary adjustments that are to be made to the cost analysis information should be done in this order:

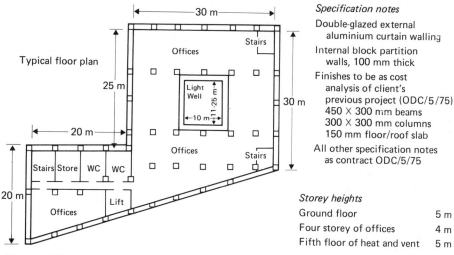

Specification notes

Double-glazed external aluminium curtain walling

Internal block partition walls, 100 mm thick

Finishes to be as cost analysis of client's previous project (ODC/5/75)
450 × 300 mm beams
300 × 300 mm columns
150 mm floor/roof slab

All other specification notes as contract ODC/5/75

Storey heights

Ground floor	5 m
Four storey of offices	4 m
Fifth floor of heat and vent	5 m

Figure 19

1 Price level
2 Quantity
3 Quality (or standards and finishes)

There now follows, complete with explanatory notes, the calculations for the practical example 2.

Practical example 2: cost plan (see Figure 19)

Element number 1: substructure

The proposed building is to be six storeys high, compared with the five storeys of the cost analysis building. However, despite this, the structural engineer's report confirms that a similar type of foundation system to that of the cost analysis building may be used, namely, a mass concrete raft with concrete column bases. The major differences between the two projects are as follows:

1 The two sites are very different from the point of view of type of soil. The cost analysis site comprises very heavy clay. The substructures were programmed to begin during October which meant that the contractor would be excavating during the winter period – it was felt that the contractor loaded the element heavily to cover any anticipated cost incurred due to the combination of clay and bad weather. The cost planned building is on a much better site with a light clay soil – the substructures are expected to be commenced in August.

2 The cost analysis building does not incorporate a complex system of ducts in the floor, therefore, it is thought that the cost of this element can be reduced considerably.

Proposed office accommodation for the Office Development Co Cost plan summary		
Gross floor area 5838 m^2	25 May 1981	
	£	£
1 Substructure	38,325	
		38,325
2 Superstructure		
2A Frame	281,158	
2B Upper floors	92,013	
2C Roof	53,145	
2D Stairs	25,944	
2E External walls	982,614	
2F Windows and external doors	7,605	
2G Internal walls and partitions	53,403	
2H Internal doors	14,589	
		1,510,471
3 Internal finishes		
3A Wall finishes	22,800	
3B Floor finishes	29,722	
3C Ceiling finishes	52,693	
		105,215
4 Fittings and furnishings	42,689	
		42,689
5 Services		
5A – 5O included together	452,503	
		452,503
6 External works		
6A Site works	–	
6B Drainage	46,460	
6C External works	8,181	
6D Minor building works	–	
		54,641
		2,203,844
Preliminaries 11%		242,422
Contingencies 0.6%		13,223
		£2,459,489

Table 4

The first adjustment is for differences in price level. This is performed by using the appropriate indices as outlined in Chapter 5.

$$£31.30 \times \frac{202}{100} = \underline{£63.23}$$

The second adjustment is for differences in quantity. In the proposed building the element unit quantity for the substructure, scaled from the plans, is 1500 square metres, therefore, the adjustment consists of:

$$£63.23 \times 1500 = \underline{£94,845}$$

The final adjustment is for differences in quality. As stated above the quantity surveyor must make an allowance for the type of soil and for the deduction of ductwork. It is felt that a truer rate for the simple foundation would be £25.55. Although this seems a considerable reduction to the original rate, having considered other rates from other contracts the range for this type of foundation lies between £20.57 (for a very shallow raft) to £35.78 (for a very heavily reinforced raft). This would tend to indicate the rate £25.55 to be much more realistic. This gives a total for this element to be transferred to the cost plan summary (Table 4) of £38,325.

Element number 2A: frame
A concrete frame is to be utilized for both the cost analysis and cost planned building. The sizing of the columns and the beams are, in the main part, the same. Therefore, it is only necessary to make an allowance for the difference in the price level and quantity.

Firstly, the price adjustment:

$$£23.84 \times \frac{202}{100} = \underline{£48.16}$$

The quantity adjustment is made simply by multiplying the updated figure by the element unit quantity:

$$£48.16 \times 5838 = \underline{£281,158}$$

As no quality adjustments are necessary, the sum of £281,158 can be transferred to the cost plan summary.

Element number 2B: upper floors
The cost plan building and the cost analysis building have been examined and it has been decided to use the same type of upper floor construction in the new building as was used before. It is considered that the spans involved are similar, thus, it is not necessary to adjust the load-bearing capacity of the internal walls.

The first adjustment that is necessary is for price levels. Once again it is simply a process of multiplying the element unit rate by the updating index.

$$£9.44 \times \frac{202}{100} = \underline{£19.07}$$

The adjustment for quantity is made simply by multiplying the updated figure by the element unit quantity.

$$£19.07 \times 4825 = \underline{£92,013}$$

Adjustment for differences in specification levels, in this case, is again thought to be unnecessary, therefore, the figure of £92,013 can be inserted in the cost plan summary.

Element number 2C: roof
In the proposed building, once again, it has been decided that the roof construction of *in situ* concrete (reinforced) is sufficient for the proposed building. The process of updating is therefore the same as for the upper floors element. Firstly, adjust for the difference in price levels:

$$£17.54 \times \frac{202}{100} = \underline{£35.43}$$

The next adjustment is for differences in quantity:

$$£35.43 \times 1500 = \underline{£53,145}$$

This revised total cost should then be inserted into the cost plan summary.

Element number 2D: stairs
The cost analysis shows that the staircase in the five storey building was *in situ* reinforced concrete with a mild steel balustrade with a sheet rubber finish to the treads and risers. This specification will ideally suit the cost plan building and no adjustment will be necessary for differences in standards. Therefore, the first adjustment will be for differences in price level. The rate that is used in this particular case is the total cost:

$$£6850 \times \frac{202}{100} = \underline{£13,837}$$

The height of each staircase in the cost analysis building is 2.8 m, with two staircases in the building, this makes a total of 22.4 m height. The height of each staircase in the cost plan building varies with the storey height. The total height will be:

$$5 \text{ m} + (4 \times 4 \text{ m}) = 21 \text{ m} \times 2 \text{ no.} = 42 \text{ m}$$

The overall width of the staircase can also remain unchanged, so the only adjustment that is necessary is to adjust for the quantity:

$$£13,837 \times \frac{42.0}{22.4} = \underline{£25,944}$$

This is the total cost and can now be inserted into the cost plan summary, as no adjustment is necessary for quality.

Element number 2E: external walls
It is envisaged that a double-glazed aluminium curtain walling system will be used for the cost plan building; therefore, the basic construction will be similar for both buildings.

The ratio of the external walls to the gross floor area of the cost plan building is:

$$\frac{5225}{5838} = \underline{0.895}$$

Adjust then for the difference in price levels as before:

$$£123.47 \times \frac{202}{100} = \underline{£249.41}$$

Adjust for the quantity by multiplying the total area of external walls by the updated cost:

$$£249.41 \times 5225 = \underline{£1,303,167}$$

The ratios of external walls to gross floor area, although not needed in this particular case, can be used to compensate for differences in floor area as follows, if the element unit rates are not given.

Assume that the only information that is available is the cost per square metre of gross floor area, which is £46.30. When this figure is adjusted in the normal way for price level, it becomes £93,53.

Now to adjust for quantity, multiply this figure by the two ratios as follows:

$$\frac{0.895}{0.375} \times £93.53 = £223.22 \times 5838 = \underline{£1,303,167}$$

Then adjust for quality in the normal way.

The cost analysis building included anti-sun glass, which the client does not require in this particular case. It is therefore necessary to make an adjustment to the rate for this variation in the specification.

	£249.41
Less cost of anti-sun glass	95.86
	£153.55
Add cost of normal glass	34.51
	£188.06 × 5225 = $\underline{£982,614}$

The cost of £982,614 can now be inserted into the cost plan summary.

Compared to the early days of cost planning and building economics, the information that is available to the quantity surveyor gives him a much better opportunity to produce an accurate figure. Not only does the method used to present the cost analysis information (shown in Appendix C in this book) give a much more detailed breakdown than has been previously available, but it can also serve as a source in itself of cost information for estimators.

Element number 2F: windows and external doors
It is assumed, for there is a lack of information at this time concerning this element, that the type of doors in the cost plan building are the same as those used in the cost analysis building. In the case of this particular element, because it is composed of two separate elements, each part should be dealt with separately. Because in this case we are employing a curtain wall system, it is possible to use a combined cost for both the doors and the windows. For the external doors and windows there is an area of 100 m² which includes for the windows not taken with the curtain walling system.

The first adjustment, therefore, is for the price level:

$$£37.65 \times \frac{202}{100} = \underline{£76.05}$$

The next adjustment is for the quantity:

$$£76.05 \times 100 = \underline{£7,605}$$

Therefore, the amount to be included in the cost plan summary is £7,605.

Element number 2G: internal walls and partitions
From the general layout of the proposed building (see office sketch plans in Figure 19, on page 79), and because the open plan system was favoured by the client in the cost analysis project, it would be wise to employ the same ratio of walls to gross floor area as the cost analysis project.

From the drawings, the area of internal walls is calculated to be: 1294 m².

Once again using the element unit rate, the cost of the element can be calculated as follows:

$$£20.43 \times \frac{202}{100} = \underline{£41.27}$$

Next adjust for quantity:

$$£41.27 \times 1294 = \underline{£53,403}$$

No adjustment for quality is necessary for this element, therefore, the cost to be transferred to the cost plan summary is £53,403.

Element number 2H: internal doors
The area of the doors required in the cost plan building has been calculated

by the quantity surveyor, by inspection of the drawings as 68 m².

Once again the element unit rate can be used and the doors can be adjusted as follows:

$$£106.21 \times \frac{202}{100} = \underline{£214.54}$$

The adjustment for quantity is:

$$£214.54 \times 68 = \underline{£14, 589}$$

No adjustment for finishes and standards is necessary, therefore, this figure can be transferred direct to the cost plan summary.

Element number 3A: wall finishes
This element is comparatively uncomplicated, in so far as the type of wall finishes are used, and it has been decided by the client and the architect to retain this system of finishes for the proposed building – namely two coats of emulsion paint on 20 mm plaster to the internal partition block walls. The area of finishes is calculated at 3800 m². It is estimated that the ratio of different finishes is similar in both cases, therefore, it is possible to calculate on the basis of element unit rates.

Price adjustment:

$$£2.97 \times \frac{202}{100} = \underline{£6.00}$$

Quantity adjustment:

$$£6.00 \times 3800 = \underline{£22,800}$$

No adjustment for standard is necessary; therefore the total cost may now be carried to the cost plan summary.

Element number 3B: floor finishes
The floor finishes element can be dealt with in a similar manner. This time, with the quantity surveyor using his knowledge of building construction and making a few common sense deductions it will be assumed that the floor finishes can again be based on the total element unit rate.

Price adjustment:

$$£3.05 \times \frac{202}{100} = \underline{£6.16}$$

Quantity adjustment:

$$£6.16 \times 4825 = \underline{£29,722}$$

As the same type of materials are being used for this element as were used in the cost analysis, there will be no adjustment necessary for finishes and standards and the total cost can be carried to the cost plan summary.

Element number 3C: ceiling finishes
It has once again been decided to use the same types of finishes as for the cost analysis building, therefore, the adjustments will be straightforward.
Price adjustment:

$$£6.24 \times \frac{202}{100} = \underline{£12.60}$$

Quantity adjustment:

$$£12.60 \times 4182 = \underline{£52,693}$$

No adjustment is required for quality and therefore the total cost can be transferred.

Element number 4: fittings and furnishings
For this element the cost analysis tries to give as detailed a breakdown as possible for the total cost of the fittings and furnishings. This element tends to be rather a 'one off' item with its own individuality, however, and it is therefore rather difficult to apply this part of the cost analysis unless the cost planned building is to be of an identical nature. Ideally then, the quantity surveyor should try to obtain a schedule of the proposed fittings, and use the cost analysis information where possible to supplement this and arrive at the cost target. In the building under consideration, the items that appear in the cost analysis are also to appear in the cost plan. They can therefore be adjusted in the first instance for differences in price level.

$$£1.84 \times \frac{202}{100} = \underline{£3.72}$$

These items can now be adjusted for quantity.

$$£3.72 \times 5838 = \underline{£21,717}$$

To this total must now be added the cost of providing the items that were not included in the cost analysis building.

It was stated by the client early in the design of the building, that he would like to have included in the cost of the building a display cabinet. There are many obviously different types of cabinets on the market, all with widely differing qualities and prices. An allowance was made in the early stages of the cost plan for this cabinet, an allowance that was thought would provide a reasonable unit. It may be suggested that this portion of the total cost of the building could be fairly flexible and if at the end of the cost planning stage it was found that there was a surplus of capital, then this could be used in providing better quality fittings. It should be noted that the reverse process must not take place, or the client would finish with a building that was not up to the standard required by the client.

The allowance that was included in the cost plan figure at the feasibility stage was £20,972 and this figure can be added to the previous figure to give a total cost to be transferred to the cost plan summary of £42,689.

Element number 5: services
It has been decided by the quantity surveyor that in the absence of any detailed information, the adjustment for the services element should be done on a cost per square metre of gross floor area. This is not the ideal solution to the problem but under the circumstances, it is the best with the information available.

It should be remembered that during the early stages of the design, the client instructed the quantity surveyor that he intended to install the heating equipment himself as he had fortunately purchased sufficient convector heaters to complete the heating element of the contract. It is therefore necessary to deduct this amount from the cost per square metre of gross floor area.

The first adjustment to make on the total cost is therefore for price:

$$£38.37 \times \frac{202}{100} = \underline{£77.51}$$

Then adjust for quantity:

$$£77.51 \times 5838 = \underline{£452,503}$$

This figure is then transferred to the cost plan summary.

Element number 6A: site works
The cost plan building has been designed to fit into a city centre site with its boundaries being the adjacent buildings. It is therefore assumed that the cost of this element can be omitted completely from the cost plan building as it will not be necessary to provide either paving or planting to any significant amount.

Element number 6B: drainage
The sketch plans at this stage will not show details of the drainage system; indeed many projects are fortunate if there are details of this element at the bill preparation stage. It is usually the case that the drainage is measured on a provisional basis, and then re-measured when the actual work has been completed. The quantity surveyor will therefore have to make an overall appraisal of the situation and decide for himself what the target cost should be. The site is in the heart of a developed area with no anticipated difficulties, so the allowance of £23,000 will be updated and used in the cost plan. This will include the connections to the public sewer, any necessary manholes and the drain runs.

$$£23,000 \times \frac{202}{100} = \underline{£46,460}$$

This amount is then transferred to the cost plan summary.

Element number 6C: external services
For the cost plan building it is estimated that the allowance that was included in the cost analysis building, when updated for differences in price levels, will be sufficient for inclusion in the cost plan summary.

$$\pounds4050 \times \frac{202}{100} = \underline{\underline{\pounds8181}}$$

Preliminaries
The overall nature of the two projects is similar, and although the cost plan building is one storey higher than the cost analysis building and therefore it could be argued that it warrants a larger allowance in the preliminaries for scaffolding, and so on, still it has been decided to adopt the same allowance of 11 per cent which is equal to £242,422.

Contingencies
Similarly, the 0.6 per cent allowance has been adopted, which is equal to a sum of £13,223.

Total estimated cost
The total project has now been calculated at a sum of £2,459,489 which represents the design target cost. It has been decided not to include the allowance of 4 per cent for the price and design risk as was done previously in the feasibility report estimate as it is felt that at this stage the design is sufficiently well designed to avoid this type of allowance. It is also noted that during this economic climate, the allowance of 11 per cent for the preliminaries is probably excessive, therefore, there is no need for a further allowance of 4 per cent.

It is noted also that the cost plan figure is £169,489 (or 7.4 per cent) above the original target or limit cost allowance. This should be reported to the client as soon as it becomes apparent to the quantity surveyor and if possible an explanation should be given for the discrepancy. In this case it is probable that insufficient attention was given to the comparative wall to floor ratios. The wall to floor ratio in the cost plan building is 0.895, whereas in the cost analysis building it is as low as 0.375. This means that in the cost plan building there is a greater amount of wall required to enclose the same amount of floor area – therefore, the design is inefficient. Due to the cost of this element, the estimate as prepared for the feasibility report, has become inaccurate. The client will now have one of three choices:

1   To abandon the scheme completely: this would be unlikely at this late stage in the design process due to the cost of all the abortive work.
2   To reduce the specification: this would reduce the cost of the project back to the original target cost. It is here that the quantity surveyor will be able

to offer some valuable advice as it will be his task to ensure that the balance of the design is retained while reducing the overall cost of the project.

3   To accept the extra cost, obtain further finances and carry on with the design.

The quantity surveyor would await instructions from the client and then depending on whether the choice was 1, 2 or 3, the surveyor will act accordingly.

8 Cost control during the detail design stage

Despite the fact that the amount of information on the proposed building was limited, an elemental cost plan has been prepared, giving a detailed breakdown of the total cost. However, to think that the figures produced as a result of this first elemental allocation are 100 per cent accurate would be extremely arrogant of the quantity surveyor and very unfair to the client. There must be some safeguard to ensure that the amounts which were included during the scheme design stage are still realistic now that the more detailed drawings are available.

This is the point of no return as far as the design development is concerned; from now on the design team's ideas should be sufficiently crystallized to warrant no further major alterations to the type of construction, plan shape, and so on. There would be little point in establishing a cost checked figure only to find at some later date that all the careful calculations are invalidated by redesign.

The process that is used to check the cost allocation is known as *cost checking*. It should now be possible to provide detailed drawings of, for example window sections, eaves details and firm specification notes for the type of finishes that are to be used.

One of the principles of cost checking is to determine whether the cost targets, allocated to the elements at the previous stage, are realistic now that the details of the proposed design are available. This stage, more than any other in the pre-contract design sequence, is where effective team work and good communications are essential to enable the detail drawings to be prepared in the order, and at the time, when the members of the design team require them for cost checking.

The method used by the quantity surveyor to determine the probable cost of the building has been cost planning. Now that this has been successfully completed, the function of cost planning has ended and throughout the remainder of the contract, cost control is the technique used (see Table 1 on page 11) to restrain the overall cost of the project in order that the tender and final account figures do not exceed the feasibility report estimate. Once the cost plan has been produced and checked, this is for many people the end of the cost planning and cost control operations. In fact, the cost of a project is just as liable to become out of hand in the post-contract stages, as in the pre-contract stages, and for this reason the overall financial control of a building project should be a continuous operation.

The technique that is used by the quantity surveyor to perform cost checking is *approximate quantities*. Not until this point has it been possible to employ this particular method of approximate estimating, as it requires a reasonable amount of detailed information as a prerequisite. The rules for approximate quantities may vary from person to person. The following notes may be taken as a guide:

1 Measurement is loosely based on the *Standard Method of Measurement* (Sixth Edition).
2 Dimensions may be scaled from the drawings and 'rounded off' to the nearest whole number.
3 Small adjustments and labours may be ignored.
4 Omnibus rates can be calculated by the quantity surveyor, who can also prepare his own descriptions to suit his method of pricing, for example:

275 mm cavity wall, comprising one half brick skin in facings PC £96.00 per thousand, 50 mm cavity and 100 mm keyed concrete block inner skin, plastered one side with 20 mm thick two coat plaster and two coats emulsion paint.

   Cost per square metre £26.80

5 All 'normal size' window and door adjustments can be ignored. The cost of providing a lintel, plasterwork and facings to reveals, flooring in openings, and so on, will not be measured, as the non-deduction for the void will compensate for this. Windows and doors are measured as 'extra over' the wall on which they occur, for example:

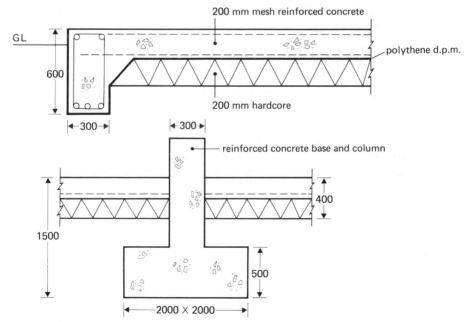

Figure 20

Extra over 275 mm cavity wall as described for 750 × 1950 × 40 mm thick solid core internal flush door, hung on a pair of 100 mm cast butts, to 130 mm × 50 mm rebated door lining, 75 mm × 25 mm softwood architrave including mortice lock and set bakelite knob furniture, and one coat primer, two undercoats, one full coat gloss both sides.

Cost per number £56.00

'Normal size' in this instance refers to standard softwood or metal windows that do not form an integral part of the actual external wall element. In the case of curtain walling, or proprietary form of double glazing, it would be advisable to measure the items in a more conventional manner, as the costs produced using the above technique would prove less reliable.

In the project under consideration, for the Office Development Company, the architect has produced his first detail drawing. It is of the element number 1: substructure, based on the information supplied by the structural engineer, and Figure 20 shows a section of the foundations from this drawing.

Practical example 3: cost checking

Cost check number 1
Element number 1: substructures

The cost check will be performed using approximate quantities, and it has been decided to split the cost check into three sections, and to calculate an omnibus rate for each item. The sections are:

1   A cost per linear metre of the edge beam
2   A cost per number of column bases complete
3   A cost per square metre for the reduced level excavation, hardcore, concrete raft and damp proof membrane

Section number 1: edge beam   Calculate a cost per linear metre of providing the edge beam, including earthwork support, excavation, levelling and compacting bottom of trenches, and so on; multiply this omnibus rate by the total length of edge beam to give the total cost of this section. The calculation is as follows:

			Cost per linear metre
Excavate trench to receive edge beam	1.00		
	0.80		
	0.20	0.16 m³ @ £2.02 per m³	0.32

			Cost per linear metre
Remove surplus spoil	1.00		
	0.30		
	0.20	0.06 m³ @ £4.14 per m³	0.25
Earth backfill		0.10 m³ @ £2.84 per m³	0.28
Earthwork support	1.00		
	0.20		
2 ×	0.20	0.40 m² @ £2.76 per m²	1.10
Level and compact	1.00		
	0.30	0.30 m² @ £0.17 per m²	0.05
Reinforced concrete in edge beam	1.00		
	0.60		
	0.30	0.18 m³ @ £50.32 per m³	9.06
Formwork	1.00		
	0.40		
	1.00		
	0.20	0.60 m² @ £8.45 per m²	5.07
Bitumen d.p.c.	1.00		
	0.60	0.60 m² @ £1.15 per m²	0.69
Polythene d.p.m.	1.00		
	0.70	0.70 m² @ £0.25 per m²	0.18
			£17.00 per m

From the drawings it has been possible to calculate the total quantity of edge beam that will be required for the building, as 178 linear metres. Therefore, the total cost of this section of the approximate estimate is calculated thus:

178 linear metres × £17.00 = £3026

Section number 2: column bases   An omnibus rate can be calculated for the cost of providing all necessary excavation, earthwork support, concrete and reinforcement, etc.

			Cost per number
Excavation	2.50		
	2.50		
	1.10	6.88 m³ @ £ 3.80 per m³	26.14
Disposal of surplus	2.00		
	2.00		
	1.10	4.40 m³ @ £ 4.55 per m³	20.02
Backfilling		2.48 m³ @ £ 3.12 per m³	7.74
Earthwork support	10.00		
	1.10	11.00 m² @ £ 2.76 per m²	30.36

			Cost per number
Formwork	8.00		
	0.50	4.00 m² @ £ 8.45 per m²	33.80
	1.20		
	0.80	0.96 m² @ £18.04 per m²	17.32
Concrete (base)	2.00		
	2.00		
	0.50	2.00 m³ @ £42.76 per m³	85.85
	0.30		
	0.30		
	0.80	0.07 m³ @ £64.00 per m³	4.48
			£225.71

It will be necessary to provide 65 column bases as this one, therefore, the total cost of section 2 will be:

65 number × £225.71 = £14,671

Section number 3: raft   An omnibus rate including for average excavation per square metre, hardcore, reinforced concrete and polythene d.p.m.

			Cost per square metre
Excavation		0.85 m³ @ £ 1.42 per m³	1.21
Hardcore		1.00 m² @ £ 1.43 per m²	1.43
Polythene d.p.m.		1.00 m² @ £ 0.23 per m²	0.23
Concrete bed	1.00		
	1.00		
	0.20	0.20 m³ @ £44.75 per m³	8.95
Formwork	178 ÷ 1447	0.12 m @ £ 2.35 per m	0.28
Mesh reinforcement		1.00 m² @ £ 1.61 per m²	1.61
			£13.71

There are 1447 square metres of raft to be provided, therefore, the total cost will be:

1447 square metres × £13.71 = £19,838

The total cost checked figure for this element, therefore, becomes:

Edge beam	£3,026
Column bases	£14,671
Raft	£19,838
	£37,535

This figure compares very favourably with the amount that was already included in the cost plan; in fact there is an overall saving on this element of £790.00. The cost checked total should now be inserted in the cost plan summary sheet (see Table 5), which has three additional columns to the one

Proposed office accommodation for the Office Development Co

Cost plan summary

Element	Initial cost target	Revised cost target	Addition	Saving
	£	£	£	£
1 Substructure	38,325	37,535	–	790
2 Superstructure				
A Frame	281,158	281,046	–	112
B Upper floors	92,013	92,119	106	–
C Roof	53,145	53,000	–	145
D Stairs	25,944	26,243	299	–
E External walls	982,614	983,067	453	–
F Windows and external doors	7,605	7,542	–	63
G Internal walls and partitions	53,403	54,276	873	–
H Internal doors	14,589	14,222	–	367
3 Internal finishes				
A Wall finishes	22,800	22,647	–	153
B Floor finishes	29,722	29,655	–	67
C Ceiling finishes	52,693	52,437	–	256
4 Fittings and furnishings	42,689	41,165	–	1,524
5 Services				
A–O inclusive	452,503	452,450	–	53
6 External works				
A Site works	–	–	–	–
B Drainage	46,460	46,230	–	230
C External works	8,181	8,436	255	–
Preliminaries	242,422	240,000	–	2,422
Contingencies	13,223	15,250	2,027	–
	£2,459,489	£2,457,320	£4,013	£6,182
				£4,013
			Overall saving	£2,169

Table 5

used previously – one for the cost checked total and a further two columns to record the net increase or decrease that this represents over the original estimate.

Cost checking is one of the tools in the quantity surveyor's hands, when trying to produce a well-balanced financial solution to the client's requirements and contain the overall cost of the building within the original estimate. As the cost of the proposed building has been allocated over its constituent elements, it is possible to establish with a quick glance whether or not one part of the project has been overloaded with capital at the expense of the others, which may result in a building with very firm foundations and external walls, but with a poor standard of finishes.

When overall cost planning and cost control techniques are used, a statement of the distribution of costs over the building is always available to the design team, allowing them to keep a firm hand on the financial reins of the proposed project.

The remainder of the elements should now be cost checked and the totals inserted in the cost plan summary. There is no special order in which the elements should be dealt with, as it is rather unwise to take any action which may be thought necessary in the reallocation of cost until the overall picture is available. It is very often done in the order that the architect produces the detailed design drawings.

As a result of performing a cost check on the complete building, it will be very seldom found that the cost checked totals for the elements exactly equal the original allowances. It follows, therefore, that in the majority of cases the cost checked total will exceed the original estimate for the element, or vice versa.

If the total cost of the cost checked building exceeds the amount that has been previously reported to the client, then in an attempt to implement a reduction in the overall cost, the possible solutions that remain open to the design team range from simple reallocation of cost, to examining alternative forms of construction and finishes. The first point to consider, in the case of overspending, is whether the amount that was included in the design risk allowance will enable the original proposals to remain unchanged. If this is not the case, then the architect will need to give thought to the redesign of certain elements, in order to produce a better balanced design. The summary of cost checked totals will greatly assist the design team in this task, as it is possible to see quickly the elements that contribute the largest proportion to the overall cost of the building, and these should be the first elements to come under scrutiny, as a small adjustment in the cost per square metre of a particularly large element could result in a large overall saving.

It is much better to try to find an alternative form of construction to solve an overspending situation, than to take what is considered by many people to be the traditional step for reducing the overall cost of the building – that is, reduce the amount that has been included for finishes, which will undoubtedly result in a building of poor quality. If the cost checked totals exceed the original estimate, it is usually due to mismanagement and poor spade work on the part of the people concerned at the outset. A step like this one, at the first opportunity, merely sends those concerned deeper into the mire, for the resultant building will not provide the client with a balanced design or good value for money – these being the objectives of cost control. If cost planning and cost control have been carried out effectively, the client and the design team may confidently expect the tender figure to be within the cost target.

9 Costs-in-use

Earlier in this book it was mentioned that one of the services that the quantity surveyor can provide for the design team is that of comparing, from the cost aspect, the economies of different design solutions. In the detail design stage particularly, when a solution to overspending may have to be sought, it may be the case of deciding on the grounds of cost, between two or more forms of construction, any of which will adequately perform the purpose for which they are intended. For example, a roof system may be provided by a proprietary form of construction, or by reinforced *in situ* concrete. To obtain a true picture of the various design alternatives, however, not only do the initial costs of a form of construction or type of installation have to be taken into account, but also the subsequent running and maintenance costs. This overall comparison of costs can be performed by considering all the costs that are associated with a building; for example:

1   The cost of the site – a purchase price (a capital sum), or an annual rent (an annual sum)
2   Cost of construction and associated professional fees (a capital sum)
3   Annual running costs, for example heating and annual maintenance (an annual sum)
4   Periodic expenditure, for example boiler replacement, renewal of roof coverings (capital sums to be met at regular intervals)
5   A further cost may be a premium paid to a landlord every xth year. Such a premium means that a lower rent is asked and, therefore, the annual equivalent is already taken into account as a deduction from the full market rent

Using techniques that will be described shortly, all the capital payments listed above are reduced to their annual equivalents, and then to this figure are added annual running costs or outgoings to find the total annual equivalent of the scheme. This method of comparison is useful to the client as it is possible to compare in this way annual liabilities with expected income.

Although the comparison of design alternatives using costs-in-use techniques is of great value, it would be naïve to think that it is the ultimate answer to forecasting cost. The disadvantages can be summarized as follows:

1   There are difficulties in obtaining information on the maintenance and running costs of various materials and systems. The Building Maintenance Cost Information Service (BMCIS) of the Royal Institution of Chartered Surveyors has been in operation now for several years, but as it is still in its infancy it needs to operate for a good many years before it can hope to show any long-term trends or economies in using particular materials. The service is similar in character and organization to the Building Cost Information Service in so far that maintenance costs are analysed along with a description of the age, construction and maintenance organization for the analysed building. Data sheets of analyses of the property occupancy costs of individual buildings are prepared from individual BMCIS subscribers as part of the reciprocal nature of membership. Occupancy cost analyses are designed to fit in with an organization's internal procedures and the analyses are used to build up a library of maintenance cost data which will help property and maintenance managers to compare their own operating circumstances and occupancy costs with those of other organizations.

2   Perhaps one of the major invalidating factors affecting forecasting techniques is inflation. Figure 21 shows how, over the last sixty years or so, the tendency has been for costs to rise. Even so, a comparatively low inflation rate, when applied to buildings with an expected life of eighty years plus, produces figures that are beyond comprehension. One can perhaps imagine a client's reaction when being told to allow £100 per m^2 for renewing felt roofing in twenty years' time.

3   The client should have a working knowledge (if not detailed) of the techniques that are to be described shortly. When dealing with the larger companies this should be no problem.

4   In a world in which many of the natural resources are gradually disappearing for ever, the prices of various materials, fuels, etc., may not just be susceptible to regular inflation rises, but in addition to increases due to demand outstripping a quickly diminishing supply. For example, the energy crisis that was brought to everyone's attention in 1973, when oil and all oil-based products rose steeply in price, will have almost certainly successfully invalidated a good many forecasts. In Britain, one result of the 1973 crisis was that the nationalized power industries, such as electricity, coal and gas, lost their heavy government subsidies.

However, despite these difficulties the techniques described below are found to be of great use to clients, particularly the larger corporate bodies, but in the light of the comments above, they must be recognized to have limitations. With this in mind, it is good practice to present a client with a costs-in-use comparison as part of a report containing qualifications about inflation and the other factors that may affect the calculations.

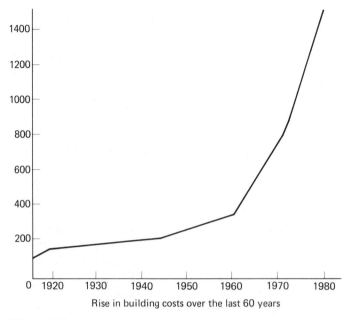

Rise in building costs over the last 60 years

Figure 21

The idea of the annual equivalent is akin to the theory of opportunity costs, as it represents the amount of interest that will be invested in property, instead of being invested at a given rate per cent of interest. It must be considered in two ways:

Annual equivalent in perpetuity

The term perpetuity in this context is used to describe assets that are not wasting away and will be in the possession of the purchaser for perpetuity, unless, of course, resold, in which case the capital is recouped – for example a freehold interest in land. On a 10 per cent basis the interest that would be earned in a twelve month period on the sum of £100 is £10. Therefore the annual equivalent of £1 invested at 10 per cent in perpetuity is:

$$\frac{£10}{100} = £0.10$$

For instance, calculate the annual equivalent, on a 10 per cent basis of £100,000, which is the price paid for a freehold interest in a plot of land.

Purchase price	£100,000.00
Annual equivalent of £1 in perpetuity at 10%	0.10
Annual equivalent	£ 10,000.00

The equivalent of a capital sum can be calculated using the following formula:

$$i = \frac{R\%}{100}$$

where, i = annual equivalent
R = rate per cent of interest

Annual equivalent of acquisitions with a limited life

To calculate the annual equivalent of acquisitions with a limited life, not only must the interest that could accrue to the capital sum be considered, but because the capital is wasting away, it is necessary to invest each year a sum, paid out of income, into a *sinking fund*. These annual sinking fund instalments accumulate at compound interest over the life of the assets to a sum that equals in amount the original capital sum invested. Thus, by the time the original capital has wasted away, a fresh sum of the same amount is ready to replace it. Therefore, if instead of investing the sum of £100,000 mentioned earlier in a freehold interest in land, it had been invested in a property with an expected life of forty years, then in addition to calculating the annual equivalent as before, a further allowance will have to be made to replace the capital in forty years' time.

The annual sinking fund allowance to replace £1 in ten years at 10 per cent can be calculated using the following formula:

$$\frac{i}{A-1}$$

where, $i = \frac{R\%}{100} = 0.10$
$A = (1+i)^n$
n = number of years

$$\frac{0.10}{(1+0.10)^{10}-1}$$
$$\frac{0.10}{(1.10)^{10}-1}$$
$$\frac{0.10}{2.59-1}$$
$$\frac{0.10}{1.59} = \underline{0.0629}$$

For accuracy, the annual sinking fund instalment must be adjusted to take account of the effects of income tax, which otherwise would prevent the annual sinking fund from accumulating to the full amount required. This can be done as illustrated below:

$$0.0629 \times \frac{100}{100-P}$$

where, P = the client's rate of income tax

It is also arguable that some allowance should be made for inflation to protect the real value of the capital assets in terms of purchasing power. To calculate the annual equivalent of £100,000 which has been invested in an asset with an expected life of ten years, the annual equivalent calculated previously, 0.10, must be added to the annual sinking fund allowance of 0.0629 to give the total of 0.1129. To avoid clouding the issue, the effect of income tax on the annual sinking fund has not been taken into account.

$$\begin{array}{r} \text{£100,000} \\ \times \quad 0.1129 \\ \hline \text{£11,290} \end{array}$$

Annual equivalent

Other formulae that the quantity surveyor should familiarize himself with are:

The amount of £1 formula
Using this formula it is possible to calculate the amount of compound interest that will accrue to a single initial deposit of £1 over a number of years. No subsequent deposits are made. The formula gives the total of the interest and the initial deposit.

$$A = (£1 + i)^n$$

where, $i = R\%$ as a decimal of £1
$n =$ the number of years over which the sum is to be invested
$A =$ the amount of £1

The present value of £1 formula
This formula can be used to calculate the sum that must be invested now to accumulate £1 at the end of a stated period, at a given rate per cent.

$$V = \frac{1}{(1 + i)^n}$$

where, $V =$ the amount needed to come to £1
$n =$ the number of years

Figures calculated using this formula will be used later in this chapter when calculating costs-in-use. It should be noted that this formula is the reciprocal of the amount of £1 formula.

The amount of £1 per annum formula
This table gives the sum which a series of instalments of £1, made at the end of each year, will amount to by the end of the term, at compound interest. The formula is:

$$\frac{(1 + i)^n - 1}{i}$$

where, $i = R\%$, as before
$n =$ the number of years

Costs-in-use example

The architect has identified three types of window that he feels will suit the functional requirements of the building. He has asked you, as the quantity surveyor, to provide him with some cost information to enable him to recommend an economic design solution to the client. This will involve the quantity surveyor undertaking a costs-in-use exercise on the three alternative window types.

The alternative window types A, B and C are designed to be incorporated into a building with an economic life of forty years, each type being subject to the following conditions:

Type A: softwood
Initial cost £6500
Replacement every 20 years at a cost of £6500
Repair cost every 10 years at a cost of £350
Redecoration every 5 years at a cost of £150

Type B: hardwood
Initial cost £7000
Replacement every 30 years at a cost of £9000
Repair cost every 15 years at a cost of £200
Redecoration every 5 years at a cost of £75

Type C: metal
Initial cost £8000
Expected life in excess of 40 years
Repair costs every 20 years at a cost of £750
No redecoration

The discount rate expected to prevail, is taken as 13 per cent.

Note: It is assumed that the cost of repairs will not include for any redecoration.

	A	B	C
Initial costs	6500	7000	8000
Replacement costs			
Type A: year 20 – £6500 × 0.087	566	–	–
Type B: year 30 – £9000 × 0.026	–	234	–
Repair costs			
Type A: years 10, 30 – £350 × (0.295 + 0.026)	112	–	–
Type B: year 15 – £200 × 0.160	–	32	–
Type C: year 20 – £750 × 0.087	–	–	65
Redecoration costs			
Type A: years 5, 10, 15, 25, 30, 35 – £150 × (0.534 + 0.295 + 0.160 + 0.047 + 0.026 + 0.014)	163	–	–
Type B: years 5, 10, 15, 20, 25, 35 – £75 × (0.534 + 0.295 + 0.160 + 0.087 + 0.047 + 0.014)	–	86	–
Equivalent capital values	£7341	£7352	£8065

The quantity surveyor would report these figures to the client with his recommendation, which in this case would probably be to design the windows using hardwood for aesthetic reasons as the equivalent capital value is within £11 of the softwood-type window.

10 Methods of valuation of landed property

Methods of valuation

Now that we have endeavoured to define the term 'value' and to equate the factors that influence it, the next step is to examine the methods of valuation available to fix a price of a piece of land, or property. The most commonly used methods are:

1　Residual method
2　Investment method
3　Comparative method
4　Profits method
5　Contractor's method

As with any technical process, there are certain basic principles that, in this case, the valuer should be familiar with, before he is able to confidently try to establish the value of a piece of property. The terms and formulae required in the valuation process are explained in the following paragraphs. The information that is included in this chapter will enable the quantity surveyor to calculate, say, the *present value* of £1, or the *amount* of £1 should the need arise. However, in practice it is not necessary to perform such lengthy calculations for each problem, as reference to a set of valuation tables, such as *Parry's Valuation Tables*, that contain all the data required in tabulated form, is generally the easiest and most reliable source of information. See Appendix A for an example of valuation tables.

The present value of £1 per annum, or years purchase

If a freehold interest in land were bought for £10,000 by a client who requires a 15 per cent return on his investment, then this can only be achieved if the interest that he receives yearly is:

$$£10,000 \times \frac{15}{100} = \underline{£1500}$$

(capital value)　　　　　　　*(net income)*

By using this information in reverse it is possible to calculate the capital value when the other pieces of information are known.

$$£1500 \times \frac{15}{100} = \underline{£10,000}$$

The expression in this example, 100/15, is known as the *years purchase*, or the *amount of £1 per annum*. Therefore it can be said that the net income × years purchase = *capital value*, and clearly the years purchase can be calculated by dividing 100 by the amount of return on capital required, which in this case would give 6.67. The amount of return on investment will alter from one client to another and will depend upon the amount of risk involved and the interest rates that are available if the capital were invested, at less risk, elsewhere.

Example
Value a freehold interest in a property producing a net income of £20,000 per annum. The client requires a return of 14 per cent on his investment.

$$£20,000 \times \frac{100}{14} = \underline{£142,860}$$

The example above concerns investments in perpetuity, that is, a freehold interest in land, the land will be in the possession of the purchaser for perpetuity unless, of course, resold, in which case the capital is recouped. For acquisitions or investments with a limited life, for example a leasehold interest in land, it is necessary to invest each year a sum, paid out of income, into a sinking fund. These annual sinking fund instalments accumulate at compound interest over the term of the lease to a sum that equals in amount the original capital sum invested. Thus, by the time the original capital sum has wasted away, a fresh sum of the same amount is ready to replace it. When dealing with the valuation of wasting assets, dual rate valuation tables should be used, as they contain a built-in sinking fund allowance.

It should be remembered that with inflation currently running at 12–15 per cent per annum, it will not be possible to purchase real estates at today's prices in twenty years, and an allowance should be made for this, or at the very least the client should be informed that all figures quoted are based on today's costs, with no allowances for inflation.

Annual equivalent in perpetuity

On a 10 per cent basis, the interest that would be earned in a twelve month period on the sum of £100 is £10. Therefore, the annual equivalent of £1 invested at 10 per cent in perpetuity is £10 ÷ 100 = £0.10.

The idea of annual equivalent is akin to the theory of opportunity costs, as it represents the amount of interest that will be forgone if, for example, instead of being invested at a given rate per cent of interest, a sum of money is invested in property.

For example, calculate the annual equivalent on a 10 per cent basis of

£150,000 which is the price paid for the freehold interest in a plot of land.

Purchase price	£150,000
Annual equivalent of £1 in perpetuity at 10%	× 0.10
Annual equivalent	£15,000

The annual equivalent of a capital sum can be calculated using the following formula:

$$i = \frac{R\%}{100}$$

where, i = annual equivalent
R = rate per cent of interest

Annual equivalent of wasting assets

When considering 'years purchase', it was explained that when investments have a limited life it is prudent to make allowance for a sinking fund. Therefore, if instead of investing the sum of £150,000 mentioned earlier in a freehold interest in land it had been invested in a property with an expected life of forty years, then, in addition to calculating the annual equivalent as before, a further allowance will have to be made to replace the capital in forty years' time.

The annual sinking fund to replace £1 in forty years at 10 per cent can be calculated thus:

$$\frac{i}{A-1}$$

where, $i = \dfrac{R\%}{100}$
$A = (1 + i)^n$
n = number of years

$$\frac{0.10}{(1 + 0.10)^{40} - 1} = \underline{0.0023}$$

For accuracy, the annual sinking fund instalment must be adjusted to take account of the effects of income tax, which otherwise would prevent the annual sinking fund from accumulating to the full amount required. This can be done as follows:

$$\times \frac{100}{100 - P}$$

where, P = the client's rate of income tax

Therefore, to calculate the annual equivalent of £150,000 which has been invested in an asset with an expected life of forty years, the annual equivalent calculated previously, 0.10, must be added to the annual sinking fund allowance of 0.0023 to give a total of 0.1023. To avoid clouding the issue, the effect of income tax on the annual sinking fund allowance has been ignored.

$$\begin{array}{r} £150,000 \\ \times \quad 0.1023 \\ \hline \end{array}$$
Annual equivalent $\quad \underline{£15,345}$

The amount of £1

Using this formula it is possible to calculate the *amount* of compound interest that will accrue to a single initial deposit of £1 over a number of years. No subsequent deposits are made. The formula gives the total of the interest and the initial deposit.

$A = (£1 + i)^n$ $\qquad$ where, $\quad i = R\%$ as a decimal of £1
$\qquad\qquad\qquad\qquad\qquad\qquad n =$ number of years over which the sum is to be invested
$\qquad\qquad\qquad\qquad\qquad\qquad A =$ the amount of £1

The present value of £1 formula

This formula can be used to calculate the sum that must be invested now to accumulate to £1 at the end of a stated period, at a given rate per cent. The formula is:

$V = \dfrac{1}{(1 + i)^n}$ $\qquad$ where, $V =$ the amount needed to come to £1
$\qquad\qquad\qquad\qquad\qquad i = R\%$, as before
$\qquad\qquad\qquad\qquad\qquad n =$ number of years

These are the 'tools of the valuer's trade'. The following examples of the various methods of valuation will show how they are used and their inter-relationship.

The residual method of valuation

This method of valuation involves calculating the *gross development value* of a building scheme, or the market price that is expected to realize when the land has been developed and disposed of, by selling or leasing, and then deducting from this gross development value (GDV) all the costs that will be incurred during its development, including the developer's profit. The residual figure represents the amount that it is possible to pay for the land, in order that it can be developed and disposed of at a profit.

Example
How much could your client afford to bid for 5 hectares of land with planning permission for detached houses at 25 per hectare? Houses in the vicinity of a similar type realize £45,000.

	£	£

Gross development value
5 hectares @ 25 houses per hectare = 125
125 houses at £45,000 ... 5,625,000

Less

Costs of construction

	£	£
125 houses @ 95 m² each @ £240 per m²	2,850,000	
Roads and sewers (say)	900,000	
	3,750,000	
Architects', QS and consultants' fees 10%	375,000	
	4,125,000	
Interest on borrowed capital say £1,000,000 for 2 years @ 14%	299,600	
Legal, agents' and advertising fees (for sales) @ 3% of GDV	168,750	
Developer's profit 15% of GDV	843,750	
	5,437,100	5,437,100
		187,900

Sum available for purchase of site × PV £1 in 2
years @ 14%
(this approximately represents the interest on the
money borrowed to buy the site plus lawyer's
fees incurred in buying it)
£187,900 × 0.7694675 ... 153,958

Say £154,000

which is equal to £30,800 per hectare or £1,232 per plot, which represents the amount the client can afford to pay for the land in order to develop it, and make a profit. It is doubtful in this case, with the limiting factors being as they are, that the client would go ahead as he is unlikely to have sufficient funds available to purchase the land. At this stage the quantity surveyor would report his findings to the client and perhaps suggest the client revise some of his requirements as to profit, or perhaps reduce the construction costs hence providing a lower quality of finish to the proposed dwellings. It should be noted that construction costs and interest on borrowing are dependent upon the client requirement. The residual method of valuation is not so much a valuation, as an estimate of how much the client could afford to bid.

Investment method

This method is used where the property produces an income, for example a shop. The income from the investment in the shop must prove to be more profitable than investment in, say, a building society.

Example
Value the freehold interest in a shop producing a net income of £8,000 per annum. The purchaser requires a 10 per cent return on his capital. Remember from previously that:

Capital value = net income × years purchase

therefore, capital value = £8,000 × $\frac{100}{8}$ = £100,000

This example is based on investments in perpetuity; leasehold investments require an allowance for a sinking fund.

Comparative method

Perhaps the most simple of all methods, this process involves a direct comparison with similar types of properties to the one being valued in the vicinity. The price paid on the open market for comparable properties forms the basis for fixing a price.

Residential properties are, in the main, valued on this basis with additions or omissions being made from the price paid for the comparable property for such things as:

1   Rear extensions
2   Standard of decoration
3   Aspect
4   Central heating (if any), etc.

Local knowledge is also useful. For example, in Hastings, Sussex, on what is known, perhaps unkindly, as the 'Costa Geriatrica', because such a high percentage of the population and the house-buying public are over retirement age, a house with a long steep flight of steps to the front door would probably be valued lower than a house having easy access.

On a housing estate, comprising many hundreds of houses, with perhaps only five different house types, the process of comparative valuation is a relatively simple affair and indeed householders can quite easily, with a little local knowledge, value their own property. However, in more distinctive areas with many individually designed properties, the valuer would have to split up, or zone the property into meaningful units of comparison.

Profits method

This method is used for properties that have an earning capacity, for example theatres, clubs, etc. It involves establishing the gross earnings for the property and deducting from this all expenses, including profit that are likely to be incurred by the tenant. The residual figure is the amount available for rent.

Contractor's method

This method is based on the theory that the value of a building and the land on which it stands is equal to the cost of construction plus the value of the site. This statement is clearly not true, as already in a previous chapter we have seen that the value of a building is the price that people are prepared to pay for it on the open market. Therefore, the idea that this method of valuation is one favoured by contractors is not true, the only time that it is used is for properties that are rarely offered for sale on the open market, such as schools, hospitals, etc.

Appendix A

Present value of £1

Number of years (n)	Present worth of £1 payable in n years' time (at 5%)	(at 10%)
5	0.78325	0.62092
10	0.61391	0.38554
20	0.37688	0.14864
40	0.14204	0.02209
80	0.02017	0.00048

Amount of £1

Number of years (n)	Amount of £1 (at 5%)	(at 10%)
5	1.2762	1.6105
10	1.6288	2.5937
20	2.6532	6.7275
40	7.0400	45.2593
80	49.5614	2048.4002

Annual sinking fund

Number of years (n)	Annual sinking fund (at 5%)	(at 10%)
5	0.18098	0.16380
10	0.07951	0.06275
20	0.03024	0.01746
40	0.00838	0.00226
80	0.00103	0.00004

Appendix B

List of elements

1 Substructure
2 Superstructure
   A   Frame
   B   Upper floors
   C   Roof
   D   Stairs
   E   External walls
   F   Windows and external doors
   G   Internal walls and partitions
   H   Internal doors
3 Internal finishes
   A   Wall finishes
   B   Floor finishes
   C   Ceiling finishes
4 Fittings and furnishings
5 Services
   A   Sanitary appliances
   B   Services equipment
   C   Disposal installations
   D   Water installations
   E   Heat source
   F   Space heating and air treatment
   G   Ventilating system
   H   Electrical installations
   I   Gas installations
   J   Lift and conveyor installations
   K   Protective installations
   L   Communication installations
   M   Special installations
   N   Builder's work in connection with services
   O   Builder's profit and attendance on services
6 External works
   A   Site work
   B   Drainage
   C   External services
   D   Minor building works

The authors and publisher would like to thank the RICS for supplying the BCIS forms used in Appendix C.

Appendix C

Blank forms - 1b

DETAILED COST ANALYSIS

CI/SfB	
	OFFICE
BCIS Code	

Job title: OFFICES	Client: THE OFFICE DEVELOPMENT COMPANY
Location: AYLESBURY	Tender dates: (1) August 1975 (2) August 1975 Base month: July 1975

INFORMATION ON TOTAL PROJECT

Project and contract information

Project details and site conditions:

Five storey office block, 40 m x 25 m, on a level site for an owner-occupier.

Contract: (To be completed by BCIS from the Contract particulars given below)

Standard Form of Contract Private Edition with quantities, fluctuating price, selected tenders. Six tenders were issued, six received, lowest accepted.

Market conditions:

Scarcity of work in the area resulted in keen pricing.

Contract particulars:
Type of contract: Standard Form of Building Contract

Basis of tender*:

[X] Bill of quantities	Open/Selected [X] competition
[ ] Bill of approximate quantities	[ ] Negotiated
[ ] Schedule of rates	[ ] Serial
	[ ] Continuation

Contract period stipulated by client ..20.. months
Contract period offered by builder ..-.. months
Number of tenders issued ..6..
Number of tenders received ..6..

* Tick as appropriate

Cost fluctuation*		YES [X]	NO [ ]
	LABOUR	YES [ ]	NO [ ]
	MATERIALS	YES [X]	NO [ ]
Adjustments based on formula*		YES [ ]	NO [ ]

Provisional sums £ ......6,000......
Prime Cost sums £ .....25,250......
Preliminaries £ ....105,144......
Contingencies £ ......6,136......
Contract sum £ 1054,180......

* Tick as appropriate

Competitive tender list	
£	N/L
1,054,180	N
1,069,283	N
1,125,276	N
1,127,359	L
1,165,222	L
1,216,532	L

ANALYSIS OF SINGLE BUILDING

Design shape information

Accommodation and design features:

The project consists of a five storey office block size 40 m x 25 m x 17.5 m high. The new building is to accommodate administrative staff for the owners company.

Areas

Basement floors	—
Ground floor	1000 m²
Upper floors	4000 m²
Gross floor area	5000 m²
Usable area	4305 m²
Circulation area	350 m²
Ancillary area	300 m²
Internal division	45 m²
Gross floor area	5000 m²
Floor spaces not enclosed	—
Roof area	1000 m²

Functional unit .... —

$$\frac{\text{External wall area}}{\text{Gross floor area}} = \frac{1875}{5000} \quad 0.375$$

Internal cube = 15,200 m³

Storey heights

Average below ground floor .... —

at ground floor .... 2.8 m

above ground floor .... 2.8 m

Design/Shape

Percentage of gross floor area:-

a) below ground floor .... — %

b) Single-storey construction .... — %

c) Two storey construction .... — %

d) 5 * storey construction .... 100 %

e) * storey construction .... — %

* Insert number of storeys.

Brief Cost Information

- Contract sum £ 1,060,316
- Provisional sums £ 6,000
- Prime Cost sums £ 25,250
- Preliminaries £ 105,144 being 11 % } of remainder of
- Contingencies £ 6,136 being 0.6 % } contract sum
- Contract sum less contingencies £ 1,054,180

Functional unit cost { Tender £ ....

excluding external works { Base date £ ....

* Amounts for single building analysed.

Blank forms - 1d

SUMMARY OF ELEMENT COSTS

Gross internal floor area: .....5000..... m²

Tender dates: (1) .August. 1975. (2).........

Element	Preliminaries shown separately		Element unit quantity	Element unit rate £	Preliminaries apportioned amongst elements		
	Total cost of element £	Cost per m² gross floor area £			Total cost of element £	Cost per m² gross floor area £	Cost per m² gross floor area (base date) £
1 Substructure	£ 31,300	£ 6.26	1000 m²	31.30	£	£	£
2 Superstructure							
2.A Frame	119,200	23.84	5000 m²	23.84			
2.B Upper floors	36,250	7.25	3840 m²	9.44			
2.C Roof	20,350	4.07	1160 m²	17.54			
2.D Stairs	6,850	1.37	–	–			
2.E External walls	231,500	46.30	1875 m²	123.47			
2.F Windows and external doors	12,200	2.44	324 m²	37.65			
2.G Internal walls and partitions	30,650	6.13	1500 m²	20.43			
2.H Internal doors	15,400	3.08	145 m²	106.21			
Group element total	£ 472,400	£ 94.48			£	£	£
3 Internal finishes							
3.A Wall finishes	14,200	2.84	4778 m²	2.97			
3.B Floor finishes	14,650	2.93	4811 m²	3.05			
3.C Ceiling finishes	30,300	6.06	4865 m²	6.24			
Group element total	£ 59,150	£ 11.83			£	£	£
4 Fittings and furnishings	£ 9,200	£ 1.84	–	–	£	£	£

116 *Practical Cost Planning*

		£	£		
5	**Services**				
5.A	Sanitary appliances	6,200	1.24	20 No.	310.00
5.B	Services equipment	–	–	–	–
5.C	Disposal installations	6,362	1.27	–	–
5.D	Water installations	36,388	7.28	–	–
5.E	Heat source	78,957	15.79	–	–
5.F	Space heating and air treatment	39,479	7.90	–	–
5.G	Ventilating system	–	–	–	–
5.H	Electrical installations	104,500	20.90	–	–
5.I	Gas installations	–	–	–	–
5.J	Lift and conveyor installations	26,300	5.26	–	–
5.K	Protective installations	–	–	–	–
5.L	Communication installations	–	–	–	–
5.M	Special installations	4,400	0.88	–	–
5.N	Builder's work in connection with services	7,700	1.54	–	–
5.O	Builder's profit and attendance on services	–	–	–	–
	Group element total	£ 310,286	£ 62.06		
	Sub-total excluding External works, Preliminaries and Contingencies	£ 882,336	£ 176.47		
6	**External works**				
6.A	Site work	20,150	4.03		
6.B	Drainage	23,000	4.60		
6.C	External services	4,050	0.81		
6.D	Minor building works	19,500	3.90		
	Group element total	£ 66,700	£ 13.34		
	Preliminaries	£ 105,144	£ 21.03		
	TOTALS (less Contingencies)	£ 1,054,180	£ 210.84		

Appendix C   117

Blank forms - 1f

SPECIFICATION AND DESIGN NOTES

Check List	PLEASE INCLUDE BRIEF SPECIFICATION AND DESIGN NOTES TO DESCRIBE ADEQUATELY THE FORM OF CONSTRUCTION AND QUALITY OF MATERIAL SUFFICIENTLY TO EXPLAIN THE PRICES IN THE ANALYSIS.
1 SUBSTRUCTURE	Mass concrete rafts and column bases, reinforced concrete suspended ground floor slab, reinforced concrete beams and service ducts.
2 SUPERSTRUCTURE 2.A Frame 2.B Upper floors	Reinforced insitu concrete frame with steel columns and beams to tank room. Reinforced concrete suspended floors.
2.C Roof 2.C.1 Roof structure 2.C.2 Roof coverings 2.C.3 Roof drainage 2.C.4 Roof lights	Reinforced concrete roof with vapour barrier, expanded polystyrene fibreboard insulation, two layer felt finish.
2.D Stairs 2.D.1 Stair structure 2.D.2 Stair finishes 2.D.3 Stair balustrades and handrails	Reinforced concrete stairs with metal balustrades and sheet rubber finish to treads and risers.
2.E External walls	External double glazed aluminium curtain walling.
2.F Windows and external doors 2.F.1 Windows 2.F.2 External doors 2.G Internal walls and partitions 2.H Internal doors	Glazed aluminium entrance doors. Reinforced insitu concrete core walls, block partitions elsewhere. Softwood flush doors with hardwood panelled doors to conference suite.
3 INTERNAL FINISHES 3.A Wall finishes 3.B Floor finishes 3.C Ceiling finishes 3.C.1 Finishes to ceilings 3.C.2 Suspended ceilings	Emulsion painted plaster generally with hardwood panelling to conference suite. Carpet throughout except quarry tiles to WC. Suspended metal ceiling finish.
4 FITTINGS AND FURNISHINGS 4.A Fittings and furnishings 4.A.1 Fittings, fixtures and furniture 4.A.2 Soft furnishings 4.A.3 Works of art 4.A.4 Equipment	Hardwood veneered cupboards, towel dispensers, mirrors, vanity shelves, vertical PVC blinds.

5   SERVICES
5.A   Sanitary appliances
5.B   Services equipment
5.C   Disposal installations
5.C.1   Internal drainage
5.C.2   Refuse disposal
5.D   Water installations
5.D.1   Mains supply
5.D.2   Cold water service
5.D.3   Hot water service
5.D.4   Steam and condensate
5.E   Heat source
5.F   Space heating and air treatment
5.F.1   Water and/or steam (heating only)
5.F.2   Ducted warm air (heating only)
5.F.3   Electricity (heating only)
5.F.4   Local heating
5.F.5   Other heating systems
5.F.6   Heating with ventilation (air heated locally)
5.F.7   Heating with ventilation (air heated centrally)
5.F.8   Heating with cooling (air heated locally)
5.F.9   Heating with cooling (air heated centrally)
5.G   Ventilating systems
5.H   Electrical installations
5.H.1   Electric source and mains
5.H.2   Electric power supplies
5.H.3   Electric lighting
5.H.4   Electric lighting fittings
5.I   Gas installation
5.J   Lift and conveyor installations
5.J.1   Lifts and hoists
5.J.2   Escalators
5.J.3   Conveyors
5.K   Protective installations
5.K.1   Sprinkler installations
5.K.2   Fire-fighting installations
5.K.3   Lightning protection
5.L   Communication installations
5.M   Special installations
5.N   Builder's work in connection with services
5.O   Builder's profit and attendance on services

6   EXTERNAL WORKS
6.A   Site works
6.A.1   Site preparation
6.A.2   Surface treatment
6.A.3   Site enclosure and division
6.A.4   Fittings and furniture
6.B   Drainage
6.C   External services
6.C.1   Water mains
6.C.2   Fire mains
6.C.3   Heating mains
6.C.4   Gas mains
6.C.5   Electric mains
6.C.6   Site lighting
6.C.7   Other mains and services
6.C.8   Builder's work in connection with external services
6.C.9   Builder's profit and attendance on external services
6.D   Minor building work
6.D.1   Ancillary buildings
6.D.2   Alterations to existing buildings

PRELIMINARIES

Passenger lift in one bank, one goods/passenger/firemens lift.

Electricity substation.

Drawings: Drawings should accompany the Detailed Cost Analysis

Index